Up
Among
the
Mountain
Gorillas

Up Among the Mountain Gorillas

Walter Baumgartel

Hawthorn Books, Inc./*New York*
Robert Hale/*London*

Library of Congress Catalog Card Number: 75-39119

ISBN: 0-8015-3097-0 American
ISBN: 0-7091-6107-7 British

1 2 3 4 5 6 7 8 9 10

Contents

Acknowledgments

This book would never have appeared in print without the help of many friends. I would like to thank the late Signora Maria Comberti, in whose house in Florence the first draft of this book, then called *Of Apes and Men*, was written, and who every evening reviewed my day's labors; Miss L. F. D. Prior Ward of Haywards, Sussex, for touching up the author's English; Peter Comberti of London, for finding my literary agent, Innes Rose, who saw possibilities in the manuscript, advised revisions, and offered it to his American colleague, Oliver J. Swan of New York, a literary agent who believed in the book and recommended it to Elizabeth Backman, senior editor at Hawthorn Books. Ms. Backman was responsible for Hawthorn's decision to publish it.

First and foremost I would like to express my deep gratitude to Dian Fossey for her warm friendship and her continuous correspondence, which kept me informed on what was happening to my friends at Travellers Rest. She has graciously granted me permission to reprint extracts from her letters, which appear in the epilogue and postscript.

Thanks also to Mrs. R. S. Van de Woestyne of Cincinnati, Ohio, who labored repeatedly through the manuscript and whose suggestions on tightening the narration deserve special appreciation.

Acknowledgments

I would also like to thank Jay Matternes, the naturalist painter, for contributing his memorable photographs of the Ugandan countryside, the gorillas in Dian Fossey's camp, and the unusual portrait of the gorilla that was selected for the jacket.

Gratitude, for which I cannot find adequate words, go to Mrs. Elizabeth Freund, whose driving spirit, indefatigable optimism, and enthusiasm are the reasons that the book came to fruition, and who prevented the manuscript from being destroyed by its discouraged author.

* * * * * * * * *

The editor would like to thank Carol Paradis for her assistance in editing and reshaping the English edition of the manuscript, and also Ellane Hoose and Naomi Ornest who spent many hours preparing the manuscript for publication.

Up Among the Mountain Gorillas

1

The Lord
of the Forest

His first warning was an angry staccato from behind a bush, followed by a second shorter, but definitely menacing, bark. The third one was raucous and reverberating, more like a scream. Everyone in our party stopped dead. And, as branches cracked, a weighty black bulk, the old silverback, hurled himself down from a tree where he had been watching us. A heavy thud! Silence!

At this moment one began to comprehend the unleashed rage and the potentially devastating strength of a 500-pound male primate defending his family.

This was the way I met Saza Chief, the biggest and mightiest of our mountain gorillas.

Another scream, a roar, and then the notorious gorilla drumroll. The invariable first reaction was: Run for your life! But leaden legs refused to obey. Again the devastating scream, the sinister drumming now in quick alternation. We flinched involuntarily under the wild rhythm. Would that long hairy arm tear apart the curtain of greenery and grab one of us? The tension was unbearable. Come what may, I would rather have seen what I was up against than to wait longer in suspense.

I got up and ever so cautiously bent the leafage aside. There he stood, the black colossus, the primeval monster, the personification of brute force. His back was silver white; his chest and shoulders immense. He turned his massive head sharply to the left and to the right, then looked me straight in the eye. His eyes were bloodshot, his wrath tremendous. I couldn't help feeling a little ashamed under his look: It did seem indecent to infringe upon his privacy.

He was so alarmingly human! I wanted to stretch out my hand and say, "Don't get so excited, old chap! Don't you recognize me? I'm your cousin!"

Reuben, my guide, did not leave us time to ponder. Crawling nearer, he beckoned to us to follow. Many a visitor has stood petrified at this point—with good reason. But those paralyzed by fear always missed a grand spectacle. Reuben well knew how far he could go.

Now the giant opened his mouth wide, as if yawning. I could have counted his teeth, and was amazed at the size of the predatory canines, not at all like humans'. From the cavity of this appalling mouth issued another hoarse, shrill, piercing sound. I gripped Reuben's leg, reassured by his closeness. I felt safer then as I watched the raging drummer belaboring chest and abdomen with huge cupped palms, in truly intimidating rhythm. No one dared breathe.

But old silverback always overdid it; he "tore his passion to tatters." At this point we began to doubt the sincerity of his wrath and to suspect that he was putting on an act. We ventured to crawl still nearer.

There he stood, the Lord of the Forest, erect in his full height, a giant of over six feet, raising his long arms. Each hand clutched a tree and shook it in savage fury. He watched our response. But the spell was broken. We stood fast and stared, seemingly unperturbed. The old blusterer was flabbergasted! With one movement he tore down branches, snapping them like matchsticks, and threw the pieces aside with such impatience that we no longer could be sure that he was only pretending.

"Now he will!" I thought, and I dug my fingers deeper into Reuben's leg. A few more screams, some halfhearted drumbeats! But Reuben's cool daring had conquered the giant. All passion spent, he disappeared into the forest, barking and grumbling.

We relaxed, wondering if Saza Chief really hadn't been enjoying this scene himself, like a tragedian of the old school, carried away by the whirlwind of his passion.

But suddenly, from below, a deep bark sounded. Were there more gorillas all around us? Terror-stricken, I looked at Reuben, who smiled knowingly. Now we are in for it! Had shrewd silverback secretly outflanked us, to divert us from following his family?

A wasted effort, for, like these much-misunderstood gorillas, we too are peaceful creatures and never follow our quarry, but retreat homeward, to savor the thrilling experience.

2

Return to Africa

It had taken a long, circuitous journey through various continents, and a good part of my life, before the day arrived in 1955 when I was able to stand face-to-face with that magnificent silver back gorilla. A number of reasons were behind this unforgettable confrontation. Not the least of these was that I had bought the inn at the Uganda border near the gorillas' habitat in the Virunga Mountains, which had become the unofficial headquarters for scientists and other visitors intent upon observing these fabled apes.

Until this period of my life, however, an adventure involving a tête-a-tête with a quarter-ton gorilla had not been what I would have considered a primary goal. In fact, it was almost by accident that I tumbled into the career that was to prove the most interesting and rewarding in my life.

German by birth, I was no stranger to Africa, having lived for many years in Cape Province and the Transvaal. I had visited Rhodesia and other countries, and during World War II had flown over the deserts of the North as an aerial photographer for the South African Air Force. After the war, when I returned to the book business in Johannesburg, which I had built up with a friend, the day-to-day routine became too boring for my restless spirit. I sold my interest to my partner and decided to return to Europe for good.

I made the fateful error, however, of traveling back via East Africa. I went by ship to Dar es Salaam, visited Zanzibar, toured the game parks and lakes, climbed Mt. Kilimanjaro, and traveled down the Nile from Uganda through the Sudan to Egypt. This trip, instead of curing me of my addiction to the Dark Continent, made me fall more deeply in love with it.

Hard as I tried, I could not settle down anywhere in Europe. I did not seem to fit in any more; I had outgrown the Western world. In this condition, I wandered into the reading room of Barclay's Bank on Cockspur Street in London one miserable November day in 1954. While thumbing aimlessly through Commonwealth newspapers, an advertisement in the *East African Standard*, Nairobi's largest daily, caught my eye:

PARTNER WANTED FOR HOTEL PROJECT IN WESTERN UGANDA

located at a junction of three crossroads, at borders of Uganda, Rwanda and Belgian Congo, at foot of Virunga Volcanoes, near Gorilla Sanctuary . . .

I was tempted. An inn located at a crossroads where the boundaries of Rwanda, Uganda, and the Congo met should be a sound business venture—but the real magnet was the gorillas. These great apes had caught my imagination ever since I had seen Bobby, the famous gorilla at the Berlin Zoo, when I was a child. Meeting gorillas in the wild, I told myself, would certainly offer the adventure my jaded spirits needed. At any rate, it would do no harm to go to Uganda and look the situation over.

My journey back was leisurely enough to avoid the sudden leap from London fog to African sunshine. A German cargo ship took me from Antwerp to Mombasa, where I boarded a surprisingly modern train and was sped to Nairobi. There I bought a second-hand Hillman Minx. It was wonderful to drive through Africa again—down the escarpment into the Great Rift, up into the White Highlands, through the tea gardens of Kericho, and down to Kismu on the shores of Lake Victoria.

I spent the next night in Kampala, which by then had become the highly sophisticated capital of Uganda, boasting large hotels, nightclubs, and cinemas, as well as a museum, stadium, theater, and a fine university. But Kampala was not the Africa I had come to find and I pressed on toward Kabale, 250 miles away.

As I neared my goal, I thought of the way in which distances had shrunk since Miss Hornby, the legendary British missionary, journeyed from Mbarara to Kisoro soon after World War I, walking with indomitable courage through the wilderness, where leopards and elephants prowled. Here I was, making the trip from Kampala to Kisoro—about 300 miles—in a day and a half!

The last lap of my safari was a steep climb over mountain roads with sharp bends and tricky hairpin turns. *Kwenda pole pole*—go slowly slowly—warned the roadside signs. My excitement mounted. Through papyrus swamps, bamboo forests, and terraced hills the narrow road twisted upward to the Kanaba Gap, an altitude of 8,000 feet. From this vantage point I could survey, for the first time, the realm that was to become my home for the next fifteen years.

The sun was just rising on that March morning in 1955 when I looked down for the first time on the wide lava plain of Kisoro, 2,000 feet below. The day was fresh and, as the mists rose from the valleys, the Mountains of the Moon seemed to appear before my eyes. Once that peaceful valley had been a volcanic inferno. The conic hills, like an ocean of frozen lava bubbles, each with a miniature crater in its truncated top, bore witness to the appalling forces that had created them.

Now all was quiet and lovely. Nature had covered the once-gray lava soil with a carpet of living green. Crops of millet, peas, beans, and sweet potatoes were growing on the steepest slopes; even inside the craters not an inch of the precious soil was wasted, and they were cultivated down to their very depths. Men and women were working in the fields, while smoke curled up from their round thatched huts. Fishermen were returning home with the night's catch from those enchanted lakes, which caught the morning light like silvery mirrors. These were the people who were to become my neighbors, many of them my friends.

Beyond this peace—lo! the Virunga Volcanoes—a fierce, fantastic backdrop, sharply outlined against the morning sky. The panorama, drawn on a gigantic scale, was overpowering. There stood the perfect pyramid of Mt. Muhavura, and joined to it by a saddle, Mt. Gahinga, shaped like a Christmas pudding and wreathed with tree heather at its crater brim. There stood rugged Mt. Sabinyo and flat-topped Mt. Visoke. Mt. Karisimbi, almost as high as Mont Blanc, towered above its neighbors, and Mt. Mikeno, 14,553 feet high and the most spectacular of the group, offered as bold a challenge to climbers as any Swiss peak. These, I knew, were long-dead volcanoes. To the west, however, loomed the "cooking pots" of Mt. Nyamulagira and Mt. Nyiragongo, still boiling, still smoking, threatening to erupt again with dramatic ferocity. These impregnable strongholds were an appropriate setting for the fabulous East African apes, the rare mountain gorillas of Uganda, who lived there. This was the Africa I had wanted to find.

I was filled with exaltation as I left the pass and made my way down to the inn called Travellers Rest. Full of confidence, eager to start my new adventure, I was intensely happy to have come back to Africa.

3

First the Landlord, Then the Guests

My spirits, so exalted on the mountains, were shortly to take a nose dive. When at last I reached the inn and met my prospective partner, I discovered that he was a wild Irish visionary, living in a dream world, waiting for a fairy godfather to arrive and pour treasure into the empty coffers of his establishment.

And what an establishment it was! The hostelry, so inappropriately named Travellers Rest, consisted of one substantial dwelling that, it was quickly pointed out to me, was the private residence of the owners, and a group of miserable huts equipped with shoddy double-deck bunks, which were meant for the tourists. First the landlord, then the guests, seemed to be the motto of this inn.

Even in my disappointment, however, I saw that the place did indeed have possibilities. The main road from Uganda branched at Kisoro, with one artery leading to the (then) Kingdom of Rwanda, the other branch going straight into the Belgian Congo. Who could ask for a more strategic crossroads? Certainly, an inn located opposite the Uganda Customs Post, where all traffic was forced to halt, had excellent prospects.

Unfortunately, the owners were not prepared to be content with anything less than a "grand hotel," with a clientele made up chiefly of millionaires. Apparently I had been expected to provide the wherewithal to make their dreams of grandeur come true. The daily quarrels that marked our relationship almost always were stimulated by my refusal to lay out my own money for overdue bills, overdue wages, and the kind of luxuries that they considered their princely right.

Even before my arrival, they had anticipated my largesse by building an "airport." How else could American millionaires come to Kisoro? I would have tried my best to talk them out of the scheme, but the damage had been done before I arrived. With an army of natives, and at considerable cost, the ground had been cleared, hundreds of lava rocks removed, and a runway leveled. They had written to all East African air services announcing the achievement and, after lengthy correspondence, one enterprising firm had agreed to give the new runway a trial.

I was there when the event took place. A multitude of tribesmen, who had never seen a "big bird" on the ground, assembled to watch the spectacle. We waited for hours. At last the tiny plane buzzed across the mountains. It had to circle many times before even trying to land because the cattle, sheep, and goats, which had been driven off the landing strip that morning, had returned and now refused to leave their pasture. Chased off from one side, they stubbornly returned to the other. Eventually we dared to spread a white sheet on the ground, announcing "all clear." The pilot, who obviously did not like the look of things, made several trial approaches but swept up again each time. At last he succeeded in setting down his craft and bringing it to a halt without breaking his neck.

"Never again" were his first words when he landed.

After a short luncheon at Travellers Rest, we took him back to the runway—and found it completely deserted. Where was the crowd?

"What's the matter?" I asked one of the few spectators who had remained. "Aren't you curious to see the big bird take off and fly away?"

"Sure we are," the man agreed, "but not from so close by. We are not as stupid as you Europeans think. To push your cars out of the mud and halfway down into the Congo, well, that is a daily routine. But to shove that thing up into the air? Nothing doing, bwana, not even for Jesus Christ."

I was relieved later when that precarious landing strip was declared dangerous. For years after, goats, sheep, and cattle again grazed there undisturbed.

My new partner's extravagance threatened to ruin me. His garage bills, in particular, were a constant thorn in my flesh. My own car had developed defects and, consequently, I had to sell it. The inn's old Ford, I thought, would suffice for us both. But I had reckoned without my partner. Returning from Kampala one night, he surprised me with an enormous grey shape parked outside in the dark.

"What on earth is this? A battleship?" I asked.

To "impress the millionaires," he had exchanged the Ford for a vintage Hudson without consulting me, but at my expense. The Hudson was an impressive sight, no doubt, but its gas consumption was prohibitive and it would never start without being pushed; even changing gears required both hands. I was very angry.

"To make up for the lost landing strip," was his excuse. "As the Americans cannot come by air, we shall have to fetch them by road, and Americans, you know, insist on comfort." A kind fate must have warned the millionaires, for not one turned up to venture a cruise in the "battleship."

I had made it a condition, on entering into partnership, that my colleague should raise additional funds, since his so-called assets had turned out to be nothing but unpaid bills. I did not propose to use my capital to pay the Indian merchants who had supplied building materials and who now insisted on settlement of their long-overdue accounts. Nor did I intend to pay the overdue wages of the African workers.

We fought not only about finances but about accommodations

for guests. The main bone of contention was *the house*. It was the only solid structure where, after some alterations, the millionaires could have spent a moderately comfortable night. It was, however, occupied by the innkeeper, and the memsahib refused to move out. The rondavels were good enough for the guests, she declared, but I certainly could not expect her to live in one of them!

When I threatened to leave, my partner had a brainstorm. I should accompany him to Nairobi, he suggested. There we would meet his father's lawyer. The old man in Ireland must have been a wise father, for he had refused to sink any more cash in his son's bottomless pit. With a "solid Johannesburg businessman" like me to back him, my colleague was sure he could persuade his father to invest in our enterprise. All I had to do was make a favorable impression on the lawyer. If I gained his confidence, he certainly would recommend the scheme and all would go well. The lawyer—an old friend of the family—wrote that he would be pleased to see us. It appeared that my partner was well connected and his father a man of standing. I felt reassured and thought the venture worth a try.

The journey to Nairobi in the "battleship" was fraught with expenditures (mine) and conspicuous consumption (my partner's). The Hudson dragged itself from garage to garage, requiring costly repairs. My colleague was perpetually hungry and ordered only the most expensive items on the menu. Needless to say, on this memorable trip he had not a cent in his pocket.

The meeting with the lawyer was, alas, a success, and he promised to use his considerable influence with the father in support of our project. It appeared that my partner's dream hotel was to become a reality. I was less than jubilant. I had come to realize that this couple would be an eternal millstone around my neck, and slowly but surely would squeeze me dry. In my heart I hoped that the father would not contribute.

During the return trip, I did some hard thinking, but wisely I waited until we had reached home safely before I informed my neurotic companion that not even for all the riches of Ireland could I remain his partner.

Although I had signed no agreement, I found it difficult to free myself of the unprofitable association. I was already too deeply involved. "Why not cut my losses and leave them to their own devices?" I asked myself. But why should I? I had come to love the place and felt sure that, by myself, I could make a success of it. I offered to buy out my colleague. However, his demands were fantastic. I would have been bankrupt before starting, had I agreed to them. So, for the time being, we agreed to maintain the status quo, each of us trying to wear the other one down. Needless to say, it was an unhappy period for both of us.

4

Early Gorilla Excursions

Over the summits of Muhavura, Gahinga, and Sabinyo runs the boundary between Uganda, Rwanda, and the Congo. Only the northern slopes of these mountains are in Uganda and these seventeen square miles, until recently fairly untouched by cultivation, had been known as the Gorilla Sanctuary since the 1920s. The other sides of Muhavura and Gahinga are in Rwanda and at the top of Sabinyo all three territories—Uganda, Rwanda, and the Congo—meet. Under Belgian rule the entire southern area was part of the Albert National Park and was closely guarded. It could be entered only by special permission of the park authorities in Brussels and such permission was rarely given. After the former Belgian territories were granted independence (the Congo in 1960 and Rwanda in 1962), the area was divided between the two new republics. The Congo section was included in the new Kivu National Park and the Rwandese named theirs Le Parc National des Volcans.

The thought of gorillas living high in the mountains above my inn intrigued me from the beginning. Everybody talked about these great apes, but few Europeans had ever seen them. I was eager to meet my neighbors. The problem was how to cover the seven miles to the foot of the mountains and back without transportation. Fourteen miles on foot, in addition to a hard day's scramble, was more than I could face. My partner had kept horses before my arrival. And a bwana on a bicycle was a rare sight in those days; should I risk loss of prestige and mount a horse of steel? Of course, I could have taken the battleship, but after the costly Kenya safari I had sworn never to go near her in the hope that her enthusiastic captain would follow suit and leave her in dock.

Business at the inn was slack. We hadn't had a guest for days and my spirits were at low ebb. Then, early one morning, Nasani, our headwaiter, called me from the garden with the glad tidings that a *wageni*, a stranger, had arrived. I quickly washed my hands and made myself presentable. Imagine my feelings when the longed-for guest turned out to be a parrot! Owing to immigration regulations, the bird had not been allowed to cross the border and its owner had returned to give us his pet.

That was the last straw! To escape the frustration, I borrowed the cook's bike and off I went toward the mountains. Passing through the village of Nyarusiza, I asked for a guide. I was directed to the home of Roveni Rwanzagire, a middle-aged villager who was to become my friend and constant companion.

Roveni was a Bahutu by tribe, but his finely chiseled features and lordly bearing betrayed a dash of Watutsi in his blood. Roveni is the native form of the biblical Ruben or Reuben and Rwanzagire means "heir of his mother's father." I soon adopted the practice of calling him Reuben.

Born in the shadow of Muhavura, Reuben for years had been the acknowledged guide for the region. However, few visitors had come to climb the mountains, so it had not been a lucrative job. Reuben's English was limited; his vocabulary consisted of "yes, sir," "no, sir," and "thank you, sir." His Swahili was not classic either and he always confused "*jana*" and "*kesho*"—yesterday and tomorrow—which often led to grave misunderstandings.

Once he showed me a clearing in the forest where, he said, his uncle had been bitten by a gorilla. This had happened "*jana*," he said, and his description of the scene was so graphic that I thought the wound must still be fresh. I wanted to hear more about the accident and insisted on meeting the uncle. Only then did it become clear that the fight with the gorilla had happened many years ago and the old man had long since passed away.

My Swahili was also rather sketchy and my Runja rwanda, as the local dialect is called, remained basic throughout the fifteen years I spent in the country, I'm ashamed to say. I am not good at picking up languages; I have to study them earnestly and there was neither a written grammar nor a dictionary in Runja rwanda. Yet Reuben and I understood each other. We had long conversations on our wanderings. When words failed, my experience as an actor came in handy.

There were other, younger men who knew the mountains equally well, but Reuben was primus inter pares. He was an enthusiast and inquisitive by nature, and an artist for art's sake in his sphere, a rare phenomenon among Africans.

Earl Denman, author of the book *Alone to Everest,* had used Reuben as a guide when he set out to climb all eight Virunga Volcanoes before tackling Mt. Everest. Denman was impressed by the handsome man who guided him up Muhavura and who begged to be taken along up Gahinga and Sabinyo also, which were then another guide's domain. Reuben was eager to serve as porter or cook, even without pay, so curious was he to learn what it was like on mountains that he had never climbed.

I was under the impression for some time that Denman had been the first and only explorer to climb all eight Virunga Volcanoes. Later, I found out that while others had done so, Earl was the only one to climb them barefoot. I asked Reuben whether he remembered the crazy bwana who had climbed the mountains without shoes. "Sure I do," he said, "and he didn't wear a shirt either!"

My first excursion into gorilla territory was wonderful. Although we saw no apes, it was fascinating to explore their habitat. After that, as often as my tired bones would allow, I mounted a bicycle and went with Reuben searching for gorillas and exploring every nook and corner of the area. With this new interest to occupy my mind, I thought less and less of cutting my losses and running away.

Reuben had often seen the gorillas and he knew quite a bit about their life and habits. His encounters, however, had been matters of luck rather than skillful tracking and for many days luck seemed to be against us. Sometimes the spoor would lead almost to the top of the mountains and then down again into the lower forests. Reuben had to cut a path through impenetrable thickets with his panga, a sickle knife found only in that part of Africa. Long-handled and shaped like a question mark, the *umuhoro,* as it is called, is a useful tool that serves many purposes, from sharpening pencils to splitting skulls. Lianas would trap me in treacherous nets, and, in the swamps, no matter how carefully I tried to tread in Reuben's steps. I would become mired and my patient companion often had to rescue me or recover one of my shoes. It was a special treat to slither down a slide of wet greenery made by gorillas or elephants, feet never touching the ground!

I enjoyed watching Reuben bend down to study a footprint, but even with all my enthusiasm for bushcraft, I wasn't really satisfied. I wanted to see gorillas. We found their nests and remnants of their meals, we heard branches crack and warning barks issue from the thickets, but the creatures themselves remained invisible.

During the dry season, according to Reuben, the gorillas would often move to the greener, wetter, and less windy southern slopes. I wanted to follow the spoor into Rwanda but Reuben would not let me. Law-abiding citizen that he was, he feared running afoul of the Belgians. Nothing could persuade him to set foot over that invisible and wholly unguarded border.

Returning from one of our unsuccessful excursions, I noticed that Reuben was gingerly carrying a little parcel, neatly wrapped in leaves and held together by creepers. I asked him what precious find he was taking home. With great care he opened the mysterious package. It contained gorilla droppings! It was evidence to prove to his chief that, although he had failed to show me gorillas, he had at least taken me to places where they had recently been.

Meanwhile, I began to question whether gorillas could ever be met in the wild. I grew tired of cycling uphill and of playing the exhausting game of hide-and-seek. I needed encouragement, and one day, when I least expected it, our persistence was rewarded.

We were walking up a narrow footpath along a watercourse leading to the saddle. Animals avoided this "main road" during the day, so presumably there was no need for caution. Reuben was some thirty yards ahead of me and we were shouting at each other, African fashion. Suddenly, on our right, there was a crash in the forest. Reuben stopped dead in his tracks and signaled to me to halt. A few seconds passed and then a gorilla family of three strolled leisurely out of the bamboo onto the path between us. Apparently the father did not like the look of things, for, as soon as he spied Reuben, he crossed the few inches of running water into the forest to our left. His mate followed him obediently. The youngster hesitated, however, and seemed curious to find out what was afoot. The little fellow looked first at Reuben, then at me, and was obviously puzzled by our strange appearance. We must have been the first humans he had ever seen. When he realized he was left alone with such uncanny company, his courage failed and the frightened creature turned and ran back into the forest. Mother, now missing her child, returned just in time to see his little bottom disappear. She chased him, grabbed him by the hand, gave him a slap on the backside, and dragged him over the path into the forest, where father's impatient barking could be heard.

This family scene, so touchingly human, changed my attitude toward gorillas. From that moment on, they were no longer mere animals to me; they were my relatives and there was no reason to be ashamed of the kinship.

Back at the inn, at that time, my partner had immediately become gorilla conscious and, characteristically, had wanted to exploit their commercial value. He had suggested that we put an advertisement in the *New York Herald Tribune* to tempt, with gorillas as bait, his cherished American millionaires to visit Travellers Rest.

"These gorillas will put us on the map," he had predicted.

"Yes, if there were any way to guarantee such encounters," I replied.

"Let's consult the *Mutwale* Paulo," he had countered. "I am sure he hunted gorillas in his younger days and will know how to go about it."

The *Mutwale* (a polite form of address for older gentlemen) was our *Saza* (County) Chief, a great hunter. He did indeed give sound advice. "Quite simple," he said. "Take six or eight Batwa with you. They are topnotch trackers and will drive the apes right into your guests' arms, my friends."

The Batwa are a pygmoid race, taller than true Pygmies but considerably shorter than Europeans. They are looked down upon by the other tribes, and Reuben was not enthusiastic about cooperating with "lazy, dirty dwarfs who will leave you in the lurch when in a tough spot." But I had heard white hunters praising the bushcraft of these little men, and I insisted that Reuben give them a try.

The results of their tracking were amazing. I began taking visitors up into the forests, confident of showing them gorillas. The Batwa searched out the apes, surrounded them, and then teased them out gently, as the *Mutwale* had predicted, right in front of our guests. Sometimes the apes came closer than the guests appreciated, but no harm was done, either to the gorillas or to us.

I had an uneasy feeling about the Batwa method, but my partner had had no such scruples. He had written to all the tourist agencies in the country, offering Safaris into Gorillaland, and he had been naïve enough even to inform the Game Warden in Entebbe, the Lord Protector of Uganda's wildlife, of our astounding accomplishments.

The Game Warden took a dim view of this enterprise and ordered it to be stopped. Gorillas, he said, must not be hunted in any way. He pointed out that, under the Game Ordinance, even photography was considered hunting and not to be undertaken without special permit. To hunt, he said, was defined as "including any unnecessary act toward, or in respect of any animal, calculated on or tending to disturb, infuriate or terrify that animal."

My colleague had been indignant. He had wanted to carry our fight up to the governor, but I had thought it wiser to give up. I was disappointed, of course, for I had taken a deep interest in these apes, of whose life and behavior we knew so little. I regretted the lost opportunity to observe them methodically in their natural surroundings.

To my partner, however, the disappointment had been overwhelming. It was soon afterwards that he had decided to accept my offer and sell his share of Travellers Rest to me. Thus I became the sole proprietor of the inn and, for the time being, had worries other than mourning the loss of my gorilla paradise.

5

Master of the Inn

Seven months and many quarrels after my partnership began came the turning point in my life. I watched the "battleship"—and all that it stood for—disappear in the direction of Kampala. I had talked my partner into accepting the car as part of the purchase price of the inn, and it was with immense relief that I saw the controversial gray monster being pushed down the road for the last time and out of my sight forever.

In reflection, I know that those seven months with my business partner had brought out my worst traits. He was at heart a kindly man, but that day, as I stood there on the road watching the receding car, the only feeling I experienced was relief. When the car disappeared, however, I felt lost and very lonely. I began to have misgivings that I had bitten off more than I could chew.

Here I was, the only European in the vicinity, alone in remotest Africa just below the equator, where life proceeds very differently. When I had lived in South Africa, I had grown accustomed to our winter being summer and vice versa, and to the crescent of the moon standing reversed in the sky. It confused me that here, near the equator, there was neither winter nor summer and that the same crescent moon swam like a bark on the celestial ocean; one was never sure whether the moon was on the wax or wane. Nor was I ever fully able to accept the Southern Cross as a substitute for the Great Bear, which so faithfully points to the North Star.

A friend in England once wrote, urging me not to miss the latest Marlon Brando film. A solicitor in Johannesburg strongly advised me to consult a local lawyer before signing a certain document. I had to reply to both that the nearest cinema and the nearest lawyer were in Kampala, 300 miles away. Kisoro did not even boast a barber; the old Indian who came every month from Kabale to cut my hair made the trip in a dilapidated native bus traveling at breakneck speed over dangerous mountain roads.

I often was asked whether I did not miss such necessities as a telephone, electric lights, or running water. I always answered no, without hesitation. I have never liked the telephone, and in all my days in Kisoro I never missed it. In later years, when severe volcanic activities threatened our region, a transmitter was installed at the police station, and this was made available to me for emergencies. I have never trusted electricity either, and was quite content with the paraffin lamps, whose friendly yellow light added a touch of romance to the stories with which I regaled my guests.

More vital than telephone or electricity, of course, was running water, and at Travellers Rest we had none. During most of the year, the rainwater stored in our tanks was sufficient, even for the needs of British bath addicts. But during the dry months (July through September and a short spell after Christmas), the precious liquid had to be brought by bicycle from a spring four miles away. Throughout these months, conservation was the rule, and guests often had to do without a bath. Every now and then someone would suggest that we bore our own well, no solution at all in a land where solid lava went down a thousand feet or more. The government could not afford such a project, much less a humble inn-keeper.

But as the new proprietor of Travellers Rest I had no time to ponder the shortcomings of Africa and the project; I had to act to overcome them. When I walked over my property and inspected what I had acquired, my misgivings increased. What a fool I had been not to bother about an inventory! How was I to run a hotel with three teaspoons and half a dozen horseblankets? Where was I to start?

The main building had consisted originally of a single medium-size room, which was to have served as the bar and dining room. There was a dark, inadequate kitchen tucked away somewhere at the back. The Irish hotelier had discovered, however, that his "grand hotel" required more space and comfort and so, as an after-thought, he had built a veranda around the entire structure. Screened with green wire mesh, it looked ugly and offered no pro-tection from wind and rain. Inside, the cheapest kind of construc-tion board had been used for partitions. The building was un-comfortable and altogether useless. And in order to turn the second building into a three-room guest house, I would have to break separate entrances through the twenty-inch lava wall, a major operation.

In order to wipe out all traces of the past, I began by giving the whole place a fresh coat of paint. Only red paint was available in sufficient quantity but, set off with black and white, the effect was pleasant. Black, white, red? A German, of course, people said. Nevertheless, they wished me luck.

The land that my predecessor had acquired for his hotel had been a barren football field; not a blade of grass, nor tree nor shrub had grown there. But he had had a green thumb and I was left with the beginnings of a nice garden, for which I often blessed him. I, too, loved gardening and fifteen years later, when I left Kisoro, the former football field had been turned into an oasis.

I was accustomed to digging soil with a spade and was annoyed when this European method would not work on Kisoro's lava. All would have been well had I been satisfied with removing the smaller rocks and smashing off the upper edges of the bigger ones, but I insisted on digging out the boulders. Only later did I realize that there was no end to digging in this fashion. Many wheelbarrow-loads of soil had to be brought to fill the holes we had made. Where I left the African ground intact, plants grew far better than where I

The beautiful Uganda countryside on the road to Kisoro, where Walter Baumgartel's Travellers Rest was located. It shows how completely the land is utilized for agriculture. Courtesy of Jay H. Matternes.

It is gorillas' midmorning siesta time, and members of Dian Fossey's Group 4 in Rwanda are lounging about. The youngsters have momentarily stopped their vigorous playing. The old female, Flossie, is somewhat apprehensive of the observer, who has just come into the clearing where Uncle Bert, the silverback male, had been napping. Flossie sits beside him with her new infant (his arm is lower right). The juveniles, Tiger and Cleo, crowd close to the new baby, obviously intrigued. Courtesy of Jay H. Matternes.

had tampered with it. The stony layer retained moisture longer, especially during the dry season. In the loose earth, water percolated down and was lost. I also found that the lava rocks contained certain chemical properties that acted as a kind of fertilizer. I had learned my lesson: Never try to reform Africa!

I planted a large variety of trees, both indigenous and foreign. Some I grew from seeds; others were given to me by the Forestry Department, which was experimenting with foreign genera. Thus I obtained a wide selection of conifers and soon had a small forest of firs, pines, and cypresses.

Lilies, gladioli, amaryllises and other bulbous and tuberous plants, as well as fushias, grew particularly well, but some European favorites, such as forget-me-nots and wallflowers, produced promising foliage but no flowers. I collected succulents and other typical African flora, but my gardeners scorned them as being "just African."

We had strawberries and wineberries, bananas, avocados, tree tomatoes, and mountain pawpaws, but citrus and other tropical and subtropical fruits would not grow at the high altitude of over 6,000 feet, so we had to import them from the Congo.

The garden provided enough of the common European varieties of vegetables to keep our kitchen well supplied. Herb gardening was an old hobby of mine and I imported a comprehensive selection of seeds, including chervil, basil, marjoram, tarragon, mugwort, and coriander. The English favorite—mint—had not been forgotten, of course. Lamb must not be served without mint sauce, even in Africa!

I was reassured when diners, who formerly had made a wide circle around the inn, began to make a habit of dropping in to risk a meal.

I was not an expert innkeeper, but in Africa many people dabble in fields they know nothing about, learning from their mistakes as they go along. I had tried various professions in my life; I had been a bookseller, actor, photographer, and secretary of an art school. From none of these occupations did I derive any true satisfaction. In

Kisoro, for the first time and late in life, I found a task that engrossed my whole being.

Every day brought new adventures worth far more than power or money. My inn remained undeveloped until the day I left. Lack of capital and general political insecurity in Africa hindered progress, and naturally foreign aid was available only to governments and not to proprietors of small establishments.

But even the shortage of funds had its advantages. It developed the ingenuity necessary to make something out of nothing. My carpenter, Elias, and I had to be resourceful. Not only were we our own carpenters, plumbers, glaziers, upholsterers, and decorators, we also made all our furniture on the spot, which often proved to be a tricky business.

The need to be economical and to use local materials turned out to be a blessing in disguise. Lava rocks served as solid walls—and they were available gratis from nature. Strewn about prodigiously, they had only to be picked up and carried to the building site. Bamboo proved to be an excellent all-purpose material, and Elias soon mastered the art of using it to cover walls and to make partitions, screens, bar counters, shelves, and even candlesticks and napkin rings. Some ceilings were covered with papyrus and others with skillfully plaited elephant grass.

I did receive some "foreign aid" in the form of a loan from a friend, Mrs. Brühl, the proprietor of the Kibo Hotel on Kilimanjaro. We built indoor bathrooms and lavatories, two more bedrooms, a large kitchen, and domestic amenities such as a scullery, a storeroom, and a laundry. Filled out with glass windows, the low-roofed veranda gave the place a quaint but homelike appearance. I turned it into a dining room, lounge, office, and personal bedroom.

We eventually benefited from British Foreign Aid, through which, shortly before I left, a pumping station was erected at our spring. This was a great achievement, providing large parts of the Kisoro plain with good water in abundance. Until then, the water supply had been our greatest problem.

Without running water, of course, there was no W.C. Pit latrines had to suffice. When I needed to dig a new one, I ran into serious difficulties. According to government regulations, they had to be fifteen feet deep. The British health officers (or "latrine snifters," as we used to call them behind their backs) knew that in the lava such a depth could never be reached, and they were satisfied with a reasonable depth of six or seven feet.

But things changed after Independence in 1962. Our first African health tycoon insisted on obedience to the letter of the law: fifteen feet and not an inch less! It was my first experience with the new type of official. Had I understood his hints and given him one of the tablecloths he so admired, I probably would have gotten away with the seven or eight feet we had achieved.

I mentioned my predicament to an African friend, a miner who held a blasting certificate. He was confident that a few charges of gelignite would do the trick. The result? A few miserable inches per charge. When the depth reached ten feet, the optimistic miner gave up. "Let us put the seat over it," he suggested. "In the dark the snifter cannot see the bottom, anyhow."

We had not reckoned with the thorough training the British had given its public servants. The snifter was fully conversant with Newton's law of gravity. He dropped a pebble down the pit and, watch in hand, counted the seconds of its fall. After some lengthy calculations, he triumphantly announced: "Ten feet and four inches!" He obviously was pleased to have found us out.

"But that lava is as hard as steel," I pleaded. "We cannot go one lousy inch farther down."

"If you can't go down, go up," he sneered. "All I am concerned about is my fifteen feet."

So it came about that the first thing that greeted the visitor approaching Travellers Rest was the pit latrine, erected on a platform five feet above the ground. But, with walls of split bamboo and a thatched roof, the little tower was quite attractive.

6

Tribes and Customs

I had willingly accepted the fact that time is no object in Africa, and I readily approved of the concept that while a straight line may be the shortest distance between two points, it is not always the best to follow. Sometimes it was wiser, I found, to bypass an obstacle than to try to go right through it. At first it irritated me that my African carpenter found it impossible to achieve a straight line or a right angle of 90 degrees, despite the spirit level I had given him at Christmas, but gradually my eyes even got used to that.

The natives' huts are unfamiliar to most Europeans. Built of mud and topped with thatched roofs, they are found all over Africa. I knew them in South Africa, where the Europeans called them rondavels. Housewives, as a rule, like them. "No dirty corners," they say. I, too, had lived in a number of such huts. On the other hand, no matter how one tries, one cannot place European-style furniture in the African circle effectively without wasting space. It is like trying to force a square peg into a round hole. "These damned beds! Why don't they make round ones?" shouted a friend who was trying to help me furnish my huts.

A scientist who was sent to study the impact of western civilization on the African mind tackled the problem from the psychological angle. He came to the conclusion that the difference between the African's mind and a Westerner's is basically a geometrical one. The African, being a product of his circular surroundings, has a circular mind. I often thought of the psychologist after Independence, when the African nouveau riche replaced their round huts with European-style houses. I wondered if he would think that square surroundings would convert the African from a roundhead into a squarehead, or from an African into a European.

When asked to which tribe they belonged, the natives in our district invariably called themselves Banjarwanda, which really means nothing but "the people of Rwanda." They actually belong to three entirely different tribes. All they have in common is the language, Runjarwanda.

The aborigines of the region were the pygmoid Batwa, heirs to a land they no longer possess. Then, the Bahutu arrived. They belong to the widely spread Bantu group of peoples and are therefore related to the Zulu and Basuto far away in South Africa. They were land cultivators, and as they cut down large tracts of the forests, most of the Batwa were driven westward into the wild hunting grounds of the Congo, leaving only small tribal groups behind.

Finally—nobody knows when—the giant Watutsi appeared. Of Hamitic origin, they came down from Somalia and Ethiopia with their cattle, stopping here and there on their long trek, founding powerful kingdoms, and leaving behind them such dynasties as the Kabakas of Buganda and the Omugabes of Toro, whose sovereign rights have only recently been abolished.

Like most tribes from the North, the Watutsi worshipped the cow. Their magnificent cattle, with the long lyre-shaped horns, remind one of the sacred Apis bulls of ancient Egypt, which are featured so prominently on the reliefs and buried in Pharaonic splendor in the Serapeum at Sakhara on the lower Nile.

In Central Africa, the Watusi found the Bahutu and related tribes firmly settled. However, these tillers of the soil were backward and easily subjugated. Impressive by their height alone, the lordly invaders gained further prestige with their cattle, the likes of which had never been seen in that part of Africa. The Bahutu coveted these useful animals. They were willing to work for years, cultivating fields, hewing wood, and drawing water, in order to possess a Watutsi cow. Using the cow as bribe, the newcomers soon established themselves as overlords.

Walking through his fields, the Watutsi squire used to greet his tenants with *"Wakosi,"* meaning "Well done! Thanks for working!" This has become the traditional greeting. It always amused me when passersby, seeing me working in the garden, called out an encouraging *"Wakosi!"* The original meaning of the word had long been forgotten.

"Thanks for working!" How nice it sounds! One would have thought the Watutsi were polite and gentle masters. In later years, however, the Bahutu fought bitterly for their freedom.

The Watutsi ruled in Rwanda. They had no political power across the border in Uganda. They were less mighty in our region. There even had been a fair amount of intermarriage between them and the Bahutu. People were proud if they could boast of some Hamitic blood. When the late Mutara Rudahigwa, the next to the last Mwami, or Sultan, of the Watutsi, gave a reception at Travellers Rest, everybody who counted himself even partly Mututsi (singular of Watutsi) came to honor him. Even some pure Bahutu proclaimed later that the visit of the "Incomparable" had been the happiest day of their lives. They celebrated the event with three days of drinking. After Rwanda's independence in 1962, however, which meant the downfall of the Watutsi, none of these enthusiasts would own up to having shown love and devotion to the Mwami, whom they had obviously regarded as their secret king at an earlier time.

In their snow-white togas, nonchalantly thrown over the shoulders, these elegant figures looked like senators of ancient Rome sitting there on my veranda. The average Watutsi is between six and seven feet tall and many reach fabulous heights. They are champions of the high jump, and a genuine Mututsi is supposed to be able to jump his own height. I fear that their technique, using a stone as springboard, would not be accepted at the Olympic

Games. I even became sceptical of their alleged record high jumps when I asked some Watutsi guests to jump over the customs barrier. "Child's play!" they said, but none of the supermen got up to follow my suggestion. Their excuse was reasonable. They had drunk four bottles of beer with some Benedictine chasers in between.

Nasani, our headwaiter, looked like a genuine Mututsi, although he may not have been of pure Hamitic stock. He had inherited the finely cut Semitic features, delicate wrists, narrow shoulders, and, above all, the most conspicuous attributes of the true Watutsi: protruding teeth and slender, calfless legs.

Like all scions of the noble tribe, Nasani was a gentleman. His father had left him a hundred cows and some fertile land and by local standards he had been "landed gentry," until a stroke of bad luck made him a poor man. To prevent the spread of rinderpest (East Coast plague or cattle fever) in the Protectorate, the Government ordered that all cattle be vaccinated. The vaccine, effective in other areas, failed in Kigezi. The well-meant pest control ended in the death of all the cows vaccinated. No compensation was paid.

Nasani, who had seen his future wiped out, quite naturally was bitter at times, and would drink one too many when he was in a black mood. At such times his poise and dignity abandoned him. As if seized by a fit, the slender body would fly up into the air, in an attempt to jump its own height. I was always afraid he would break his neck as he danced himself into a frenzy, head thrown back in sharp jerks and the arms, with an imaginary spear brandished in each hand, propelled madly.

Nasani was exceptionally gifted, a man of many talents. An accomplished waiter, he was loved by all the guests, especially children, who would eat better when he served them, his pranks whetting their appetite. Nasani also made curtains and cushions, repaired the linen, and even sewed on my buttons and darned my socks. He was a master in the art of folding napkins into fantastic shapes and patterns but he guarded his secrets jealously. Even I was not allowed to watch him.

7

Catering to Apes

Much water had flowed down the River Nile since the Gorilla Sanctuary had been put out of bounds for me. Respecting the warning of the Game Warden, I had not ventured into the forests, "disturbing, infuriating and terrifying" the animals. The longer my inn was established, however, the more I longed to trek up into the mountains. I thought of the gorillas especially just after sunrise, when the three summits, with every familiar detail temptingly clear, stood out in relief against the blue morning sky. Were the apes up already, collecting breakfast in their "garden," while I was cutting flowers in mine? Or would they remain in bed until a stronger sunlight pierced the canopy of leaves and warmed the damp forest? Were the lobelias, which I had seen only as withered sticks, in bloom? What about the mountain orchids I wanted to dig out for my garden? I also missed Reuben, my jolly scout. It was altogether against my grain to potter about in house and garden, while out there, almost at my doorstep, the great adventure awaited me.

When I learned that the Game Ranger from Mbarara, who was in charge of the Southern Range including the Kisoro Sanctuary, was coming to visit my inn, I welcomed the opportunity to send out feelers in the direction of Entebbe—the long-time seat of colonial government offices—and of the Game Warden. I wanted to convince him that my idea of making the gorillas available to visitors was not only sound, but perhaps essential to the crucial question of conservation.

Africa was in a state of ferment. Radical change was imminent, a change which well might spell doom for the gorilla. Under the British they were now safe. The Belgians, too, were guarding theirs like Cerberus. Would the same spirit of conservation exist under African rule? I was afraid not. The African is a realist. Reserving large tracts of arable land for animals would seem folly to him.

As long as we Europeans were responsible, it was our task, I believed, to safeguard the future of Africa's wildlife and try to make the Africans appreciate their unique heritage, which, once destroyed, could never be replaced. Time was running out. Some species, such as the rhinoceros, were already doomed to extinction. The gorilla might be next.

Impregnable as the mountain sanctuaries appeared, they were already being eroded by land-hungry natives. The governor of Uganda had yielded and ceded to them a wide strip of the original gorilla forest. Settlements were encroaching on the woodlands, which not long ago had stretched down to the base of the mountains. Huts were climbing higher and higher up the steep slopes. Trees were being felled and crops planted where the gorillas had enjoyed their Lebensraum, their richest feeding grounds. The apes were not to be routed easily; they still built their nests there, often close to the new huts. But they were bound to give way to the pressure of the already numerically superior human race.

Only when lovers of nature the world over became gorilla conscious, acquainted with their existence and their problems of survival, could the salvation of this endangered species be achieved. By education and by contact, an influential public opinion might be created, which the future rulers of the country would not be able to afford to ignore.

I presented my argument to the Game Ranger. Sympathetic, he

agreed to intercede with his superior in Entebbe, with the result that the Chief Game Warden forgave my past sins and granted me permission, under certain conditions, to observe the apes to my heart's content and even to take genuinely interested visitors on our excursions. The warden himself suggested a gorilla-feeding scheme. As an experiment, I was to plant various crops on open mountain spaces, hoping that the gorillas would be attracted to them.

It was an interesting but impractical idea. The Sanctuary was part of the larger Forest Reserve and the Forest Department rightly objected to cultivation in the area, fearing that the natives might then claim more land. They would certainly protest a European being permitted to cultivate land that had been refused to them. Furthermore, gorillas, being vegetarians, required an enormous quantity of food. It would be impossible to grow enough crops as quickly as these gluttons could harvest them. One also had to reckon with elephants and buffaloes, whose healthy appetites would devour in a jiffy what had taken months to grow.

So began our experiments in catering to apes. Instead of trying to grow crops, we decided to deposit food on paths and other places where, sooner or later, the gorillas were bound to pass. In this way we hoped to lure them into coming regularly to clearings where they could be seen better than in the dense undergrowth. Camouflaged observation posts would be erected there, and substantial meals served.

The first item on the menu was sugar cane, but our fastidious guests neve tried it. Had they sampled its inherent sweetness they probably would have found it appealing but all they did, as Reuben once observed, was play with the stringy sticks.

Sweet potatoes and bananas were offered as second and third courses. Both were unknown to our gorillas. If bananas or plantains grow near their habitations, gorillas love them, for they eat not the actual fruit but the stem, the pith of which is highly nutritious. The entire plant, which takes about two years to mature, is thus destroyed. A raid on a grove can cause a severe shortage of food for a whole village where bananas or plantains serve as the people's staple diet. We offered our banana bait in clusters. We hung them singly, like pine cones, on bushes and low trees. But in vain!

We then switched over to maize cobs. The natives had told me that when they were growing maize on the mountainsides the gorillas used to raid the shambas. When they planted English potatoes instead, the raids ceased. But our gorillas seemed to have lost their appetite for maize. No matter how skillfully we placed the cobs, they remained untouched.

Salt, I said to myself, is a commodity desired all over Africa. In the past even the natives would walk until their feet were sore in order to exchange an elephant tusk, a lion skin, or even gold for a couple of handfuls of salt. Animals travel long distances to lick salt at places known to them. Perhaps the gorillas were salt-hungry.

As a start, we carried a few bags of rock salt up to the saddle and established several salt licks. We examined them daily. A few weeks later, after the ground dampness had dissolved the salt and the buffaloes had trampled it into mud, distinctive footprints indicated that gorillas had indeed visited the spot. When their nests showed that a family had spent the night nearby, I was delighted. At last the right bait had been found! A ton of salt, gift of the enthusiastic Game Warden, was carried up the mountain. Naturally, the unpredictable apes never returned to the lick.

I wrote to a firm and asked for samples of certain mineral bricks, dietary supplements usually put out for cattle. The merchant replied: "We wonder whether, as a change of diet, your gorillas might be interested in our milk powder and condensed milk products, known to be popular with chimpanzees." I wrote back that one could perhaps make this diet just as popular with gorillas but that there was a slight difference between feeding a pet chimp or two in a circus or zoo and a whole forest of gorillas. The adult male gorilla averages six feet in height and weighs more than five hundred pounds.

The feeding scheme had cost me much time and energy but the

effort was not entirely useless. It had brought me into close daily contact with the animals and a tracking technique had been developed. We could now provide visitors with the unique experience of meeting Uganda's Mountain Gorillas on their home ground.

Furthermore, the Game Warden had grown to respect my efforts and, as a token of his confidence, awarded me the title of Honorary Game Ranger.

I had neglected my duties as innkeeper while conducting the feeding experiments. To cope efficiently with both gorillas and guests I needed help. I needed someone prepared to share life and work with me, taking turns up in the forests and down in the inn. This person would have to be an idealist, for all I could offer in return for service was free board and lodging.

Early in 1956 I approached Dr. L. S. B. Leakey, the renowned East African anthropologist and curator of the Coryndon Museum in Nairobi, and asked him if he could recommend a suitable helper, a student, perhaps, who would be interested in collecting valuable material for a thesis on gorilla behavior.

"I only wish I were young and free; I would jump at it myself," Dr. Leakey wrote back. He promised to keep the matter in mind.

8

Are There No Brave Men in East Africa?

"I have found a suitable person for you," Dr. Leakey wrote several weeks later, "provided you do not insist on having a man." I replied that, while I thought it was a man's job, I'd accept a female of the right sort. The "right sort" was Rosalie Osborn, a Scottish lass of twenty-two, then working at the Coryndon Museum in Nairobi.

Miss Osborn had had no scientific training; nevertheless, Dr. Leakey had sent her to collect fossils on Rusinga Island in Lake Victoria. This was the spot where, some years earlier, Mary Leakey, Dr. Leakey's wife, had found an important skull from the Lower Miocene period, the so-called "Proconsul." The Proconsul was, in Dr. Leakey's own cautiously chosen words, "a near approach to the form of apelike creature from which the human stem eventually was evolved." It lived twenty or more million years ago in the region of Lake Victoria. Rosalie had done remarkably well, Dr. Leakey said, unearthing additional fossils from the same period.

Rosalie had a natural gift of observation and was an alert, competent, resolute young person. Perhaps her only weakness was that she, like many people of her age, wanted to do too many things at once. Shortly before the time she was supposed to arrive, she wrote that Dr. Leakey had asked her to collect still more fossils, this time in the Gulf of Kavirondo, and Sir Mortimer Wheeler, the archeologist of BBC fame, wanted her to work for him in Tanzania. Could she postpone, for a time, the Kisoro adventure?

I had counted on my helper's early arrival and was a little angry about the proposed delay. The gorillas and I were not prepared to wait so long. I wrote back suggesting a compromise. I would concede to her the Gulf of Kavirondo. "As for the Wheeler project," I wrote, "you will have to make up your mind: either the gorillas or Sir Mortimer." I proudly report that Rosalie chose the apes.

Rosalie was not squeamish. She would rather mix cement than help in the kitchen. She more than made up for her inadequacies on the home front by the courage, determination, and enthusiasm that she showed in the forest. She did not mind camping on the saddle between Muhavura and Gahinga, accompanied only by Reuben and the trackers, a challenge even to the toughest. At 10,000 feet the nights are bitterly cold and windy, but I never heard one word of complaint from this young woman.

We were still attempting to feed the gorillas and distributing food over the wild rifted terrain was hard work. For a time, Rosalie had a spell of bad luck, for the gorillas that she had gone to observe kept themselves hidden. One day, however, while eating her lunch in the shadow of a tree, Rosalie had the uncomfortable feeling that she was being watched. Looking about, she saw at last a gorilla watching her from the crotch of a tree. Resting against the trunk, arms folded, he sat there nonchalantly looking down at her.

"A penny for your thoughts," Rosalie said. He ignored that remark and commenced to scratch himself freely in a way that a young man should not do in the presence of a young lady eating her sandwiches. Pretty soon he got down from his chair and strutted back into the forest.

Somehow this story reached the Scottish press—with irremediable consequences. Just when Rosalie was beginning to meet the gorillas more frequently, a letter from her mother arrived. Rosalie had not advised her mother of her whereabouts and unusual activities. Nevertheless, the letter stated flatly that "the feeding of baboons had to stop forthwith." Rosalie was ordered to look for a solid job in a respectable British community. "How does Mother know?" Rosalie wondered. "I send all my letters through friends in Nairobi."

I understood the implication about the respectable community and was duly offended, although, I had to admit, her mother was justified in her reaction. How could a young girl of good family live unchaperoned with a bachelor—a Continental one at that—running a pub in darkest Africa! What could the girl be thinking of! Mother threatened to take the next plane and free her child from Bluebeard's grip.

"If Mother meets you and sees the place, she will surely calm down," Rosalie said. But I balked at the idea of having any mothers around. Her parent should have known that Rosalie was safer with baboons and savages than any girl in the streets of London.

To my dismay, Rosalie proved to be an obedient daughter and left Travellers Rest for the famous Big Game Farm of Carr Hartley in Rumuruti, near Mt. Kenya. The respectable British community there consisted of some uncouth South African wild-game catchers side-by-side with many other non-British creatures, such as lions, rhinos, cheetahs, and a motherly baboon, which—such is the irony of fate—bit Rosalie fiercely on the arm on the very day of her arrival.

Later Rosalie returned to England. On the strength of her gorilla observations she won a scholarship to Cambridge to study zoology and received the academic degree she always had fervently desired. And then? Rosalie, of course, returned to Africa and is teaching in a high school in Kenya.

Her mother eventually did visit East Africa and Rosalie brought her to Travellers Rest. While Rosalie visited her old hunting grounds, her mother stayed at the inn with me. We got on very well and I told her how deeply she had hurt us—the gorillas and me—by making Rosalie leave us because we were not "respectable."

"Of course, had I known you—I can't judge the gorillas, I haven't seen any—I would not have minded," mother said. "But imagine my shock when I saw this bold headline in the *Edinburgh Evening Standard:* 'SCOTTISH LASS LUNCHES WITH YOUNG GORILLA-MAN.'

"That sounds interesting, I thought, and bought a copy. Imagine my consternation when the 'Scottish lass' turned out to be my own daughter, whom I believed to be sitting safely behind Dr. Leakey's typewriter in Nairobi!"

Rosalie had to be replaced. I did not dare bother Dr. Leakey a second time, so I put an advertisement in the *East African Standard,* the Nairobi daily that had brought me to Kisoro. It read:

THREE MONTHS FREE HOLIDAY
to lover of wildlife
willing to assist in experiment
with GORILLAS.
Unique chance for scientific observations.

I received many applications, but those who answered invariably wanted only to make a profit. The one reply that clicked was signed Jill Donisthorpe. Not being familiar with English names, I assumed that the writer was male. Our correspondence seemed to confirm that assumption. In one of my letters I described graphically a gorilla mating scene that I had just observed. Had I known that Jill was a woman, I certainly would have treated the theme more delicately. Imagine my shock when an old Standard Vanguard, held together by wire and string, stopped at my door and discharged another female!

"Good God!" I greeted her in shock. "Are there no brave *men* in East Africa?"

But Jill, more mature than Rosalie, was tough. "She is as strong as a buffalo," said one of the natives.

Newspaper advertisements had played a fateful role in Jill's life. As a young girl she had come across one in an English newspaper asking for a "woman scientist for the study of gorillas in Africa." Perhaps it was the trick of a white slave trader, Jill guessed, for no scientist, to her knowledge, was studying gorillas at that time. Nevertheless, the mysterious advertisement captured the young girl's imagination and she decided to study zoology. Her mother had warned, "What does a woman do with zoology? Become a teacher? That is the last thing the Lord had in mind when he created you." How right she was. Jill was a bird of passage, a seasoned hitchhiker, always on the move. Now a second newspaper advertisement—mine—had closed the circle and brought her to her dreamed-of gorillaland.

The gorilla-girl, as she soon came to be called, was amazingly lucky right from the start. She met gorillas almost every day, returning, usually at an early hour, with a triumphant smile and interesting experiences to report. Reuben was flabbergasted. For Rosalie, meetings had not been quite so frequent. He asked Jill what sort of religion she observed, for, in his opinion, it had to be a mighty god who sent the gorillas so easily across her path.

Reuben was a religious man, an elder of the Church Missionary Society. Every night he held a worship service in the camp, leading in prayer while his companions responded in chorus. In my hut I could clearly understand every word of his litany, which went something like this:

God the Almighty, our Father in Heaven! We thank Thee for the *posho* and beans you have give us today. Let there be more tomorrow. As for me, you know that I have a weak tummy and *posho* and beans do not agree with me. Please put it in the bwana's mind [mine!] that he offer me again of his rice and bully beef as done today. Let the sun shine for it is damned cold on this mountain when it rains. Let the spirits of Muhavura, Gahinga and Sabinyo look down upon us benevolently. Let us meet the gorillas and let them be in a friendly mood. Lord, Thy Name be praised! Amen!

This devout Christian and church elder also never failed to call upon the spirits of the mountains; double insurance could do no harm! Reuben had no fear of gorillas, but sometimes the mountains gave him the creeps. He knew that volcanoes, in particular, could never be trusted. Perhaps his uneasiness was justified. One day he came down with the news of hot earth spewing from a slope halfway up Muhavura. There was no doubt about it: Something was smoking up there. I could see it clearly through my binoculars. We were having daily earth tremors at that time and a small crack had appeared at the foot of Sabinyo. An eruption on the flank of Muhavura, therefore, was not utterly improbable.

I was on crutches, having broken my leg accidentally just before Jill's arrival, but Jill wanted to go and investigate at once. Reuben, always obliging, stubbornly refused this time to go with her. I had never seen him in such a mood. He capitulated at last, but insisted that Jill give him a written statement verifying that he, "Reuben Rwanzagire the Guide," had warned her and would not be held responsible for any mishap that might occur. When they neared the smoking area, Jill reported later, Reuben implored her "for God's sake" not to go any further. When she didn't heed his warning, he turned into a maniac. Arms raised, he sank to his knees, singing and praying like an Old Testament prophet. "Shut up!" Jill shouted at him in irritation. With that madman at her side, she suddenly no longer felt quite as safe herself on the allegedly erupting volcano.

It proved to be much ado about nothing. Reuben's hot earth was a harmless landslide and the smoke rising from it nothing but ordinary dust.

BIRDS OF JOHN BURROUGHS
Keeping a Sharp Lookout
edited and with an introduction by
Jack Kligerman
foreword by Dean Amadon
including 10 illustrations by
Louis Agassiz Fuertes

Superb collection of eleven of Burroughs's essays on birds, which demonstrate the everlasting values of nature. Some of Burroughs's finest essays, including "The Spring Procession," "Winter Neighbors," and, of course, "A Sharp Lookout," are included, accompanied, as they were at the time of their original publication, by the magnificent bird illustrations of Louis Agassiz Fuertes.

hardcover **$6.95**
paperback **$3.95**

RESCUE AND HOME CARE OF NATIVE WILDLIFE
By Rosemary K. Collett with Charlie Briggs

This is a unique and essential reference volume for any resident of a rural, suburban, or urban area who has ever encountered an orphaned or injured wild creature and felt the impulse to nurse and nurture it to self-sufficiency.

She gives detailed instructions on how to best house, feed, handle, and nurture these animals during the period of dependency resulting from orphaned infancy or disabling injury.

The author's incidental anecdotes will also prove fascinating reading for every individual interested in natural history and wildlife preservation.

paperback **$3.95**

UP AMONG THE MOUNTAIN GORILLAS
The Tales of an Innkeeper in Uganda
By Walter Baumgartel   foreword by Dian Fossey

Walter Baumgartel tells of his adventures as innkeeper of Travellers Rest in Kisoro, Uganda, at the borders of Rwanda and the Belgian Congo. During his fourteen-year proprietorship, Baumgartel's quaint hostel and thatched guest huts became the informal headquarters for gorilla observers from all over the world. This book tells about Baumgartel's personal experiences with gorillas and his adventures with other gorilla buffs. Illustrated with photographs and maps.

hardcover **$9.95**

ORDER FORM

THE WALKING ADVENTURES OF A NATURALIST
By John K. Terres
Illustrations by Charles L. Ripper

For eight years John Terres, distinguished naturalist and former editor of *Audubon,* observed by day and night the wild creatures who inhabited the swamps, ponds, and overgrown fields of Mason Farm in North Carolina. In *The Walking Adventures of a Naturalist,* which won the John Burroughs Medal for distinguished nature writing, Mr. Terres writes with a scientist's knowledge and a poet's vision of birds and mammals, making of each a living portrait and a testament to the wonders of nature.

paperback $3.95

HOW BIRDS FLY
Through the Water and through the Air
By John K. Terres

John K. Terres has devoted a lifetime to patient observation of birds and other wildlife. In this book he gives an exciting, detailed account of birds in flight, pointing out their beauty and natural grace as well as the technique and engineering that makes their flight possible.

Here, too, are facts on how fast birds fly, how high (or deep) they fly, and what distinguishes one kind of flight from another. And throughout are fascinating bird-life anecdotes drawn from Mr. Terres' years of experience as ornithologist and editor of *Audubon Magazine.* Illustrated with line drawings.

paperback $3.95

9

Gorilla Country

As soon as we had permission to resume tracking gorillas we began taking small groups of guests up to gorillaland. Those were memorable experiences for any lover of Africa, and nature, even if he failed to meet the great apes he had come to find.

We would leave the inn after an early breakfast and drive seven miles through fields and pastures to the foot of the mountain. In this green, unspoiled landscape every tree, every hut seemed to stand at just the right spot. The people, too, were just right, blending into the landscape as if God had created them out of their own black lava earth.

The men we met on the way were friendly, polite without being obsequious. Only rarely would we encounter one who failed to greet us with a jolly *"muláho muláho!"* Those of us familiar with the intricate pattern of local greetings would respond with a melodious *"mulahonéza."* The native would reply, *"namáhoro,"* and an equally melodious *"namahoronéza"* would close the courteous exchange. On the bad roads, of course, it was safer to have both hands on the wheel, but I always managed to have one hand free so that I could return each *"muláho"* by waving like a monarch.

A gang of schoolboys in neat khaki shorts and clean white shirts often hailed our car. Cheeky and bursting with vitality, they would risk standing in the middle of the road, daring the approaching car to stop and only at the very last moment jumping aside with great hullabaloo. We often met other boys in rough, old-fashioned goat-skins, driving their long-horned cattle across the road just when we wanted to pass. We passed women working in the shambas, cutting maize or millet and hoeing patches of peas and beans. They were too shy and modest to greet us at first, but my daring flatteries, drummed into me by my trackers, soon melted the ice. The ter-raced cinder cones, hills on which the women were laboring, are the most striking feature of the Kigezi scenery. They are so steep that the women can work the shambas without bending down, a blessing to the babies carried on their backs.

The lava soil of these hills, which are cultivated right to the top, appears rich and fertile, at least during the nine months of rain. It does not rain every day nor all day long, even during the wet period. In the early afternoon, there is usually a short heavy shower, sometimes a torrent, then the sun shines again. The sky seldom remains grey and overcast all day long as it does in Europe.

The porous lava soil does not retain moisture for long, and dur-ing the three rainless months it soon becomes dry. If the new rains do not set in when they should, a short hungry period ensues. Pro-gressive agricultural methods have not yet been adopted but the natives have not overworked the soil and erosion is not a major problem. In this densely populated region they cannot afford to let fields lie fallow for even one year. Their own traditional system of crop rotation, however, is effective enough: crops are cultivated alternately in the ridges and furrows; the weeds, left lying in the furrows, act as fertilizer.

There are no real cash crops in these parts. The millet, sorghum, maize, sweet potatoes, plantains, peas, and beans, grown on the plain, and the wheat and potatoes in the mountains, are cultivated mainly for subsistence. Most, if not all, of these crops were introduced during the first half of this century. One wonders what the Africans ate before colonists brought in these staples.

Our route to gorilla territory passed through Nyarusiza, the village where Reuben and his helpers lived. Here the road deteriorated into a grass track. Boulders, shaken down from Muhavura by a landslide, blocked the road in places. In heavy rains the car invariably got stuck in the mud, but people were always ready to push it out with much noise and laughter.

The sign, "Guide Reuben," indicating Reuben's residence, could not be overlooked. I painted it myself in bold letters, and Reuben and I were both very proud of it. Panga on shoulder, he was always ready when our party approached. He knew he was the world-famous gorilla guide, and would climb into the car with a smart salute.

The car had to climb a thousand feet up from the inn before this signboard, erected by the Game Warden, was observed: "Gorillas are dangerous and should not be closely approached." At this point, Reuben always smiled reassuringly at our guests. Here we would abandon the vehicle for the narrow footpath that meanders gently uphill through wheat and potato shambas. We then would follow a man-made watercourse, fed by numerous trenches dug in the swampy ground. The path grows steeper and steeper as it leads into the hagenia forest.

The altitude of the lower woodland is between 7,500 and 8,500 feet. *Agauria hagenia* and *Myrica salicifolia* grow here. Trees thirty to forty feet high, they are profusely festooned with lianas and creepers. In the bushy undergrowth, *Vernonia*, with its mauve-white heads, is the predominant shrub. This region serves as the gorillas' herb and vegetable garden, where they can feast on dock, bracken, thistles, and, most desirable of all, giant celery. Brambles grow here, too, their leaves and berries providing tidbits for the giant apes.

The governor had ceded this strip of forest to the natives and, as more of the forest was cut down, the gorillas retreated to the bamboo zone higher up. But they had not been driven away completely. Their nests were found here frequently, and Reuben almost always could tell how long ago the apes had slept in them. If the beds had been occupied the previous night, it was child's play to follow the spoor and catch up with the quarry.

Between the lower woodland and the bamboo belt we had to cross the forest demarcation line—with due caution. It was here that an elephant once snatched a gun from a game guard's arm. The bamboo zone girdles the three mountains at about 9,000 feet. A fernlike giant parsley thrives here and in the clearings glow *kniphofia,* the same red-hot poker that is cultivated in English gardens. Both are eaten by gorillas. Their daily bread, however, consists of tender young bamboo shoots, whose points pierce the forest soil throughout the year. The shoots are eaten whole, but only the inner pith is selected from the larger, older stems. The peeled-off sheaths litter the trail like discarded sandwich papers.

An expert once told me that bamboo grows twelve to fifteen inches a day, reaching its full height of eighty to one hundred feet within a couple of months. "Any fool can see it growing," he said. In all my years in the forest, however, I never succeeded in doing so. Perhaps he had watched a different variety. Our *Arundinaria alpina* does not grow as fast or as high as, for example, the long-caned species found in Cambodia. It grows so densely, though, that progress through the thickets is slow and tedious.

On rocky patches a miniature violet, *Viola abyssinia,* is found, but the nettles are so malevolent that I confess I left the flora of the bamboo belt largely undiscovered!

As the terrain became steeper and the undergrowth thicker, Reuben's *panga* never rested. Visitors would begin to feel the effects of the unaccustomed altitude and stop more often to recover breath. I often got claustrophobia as the trail led up, down, and in circles through this nightmare of thin, disorderly sticks, and I always heaved a sigh of relief when we emerged from that bamboo prison into the open, parklike landscape of the saddle.

I loved this saddle land, always fresh and green. Buffalo grazed there, and it was the realm of elephants. Streams murmured through swampy meadows. *Alchimilla cardamine,* yellow and mauve ground orchids, and many varieties of mosses, reeds, and marsh plants made the saddle a veritable paradise for the botanist.

The vista of the hills, lakes, fields, and far-distant mountains always seemed enchantingly new. On clear mornings the glacial peaks of the Ruwenzori were visible. On dark nights I could see the warm gleam of the paraffin lights at Travellers Rest.

Four thousand feet above my inn, between two mountain peaks, I always felt as if I were on another planet. The faint barking of a dog, the muffled mooing of a cow were the only links between the silent world around me and the populated plain below. The automobiles, racing with arrogant importance over the roads of Africa, looked like flimsy toys through my binoculars. Their speed seemed futile, their eagerness absurd. "Why all this hurry?" I wondered.

When I would journey by myself to the sadde, I found that in the mountains, among the animals of the forest, time did not exist: I ate when I was hungry, rose with the sun and went to bed with the sun. If caught by wind and weather, naturally I hurried back to shelter, like any other animal. But on fine days I took my ease, studying the technical wonder of a reed, admiring the intricate form of an orchid, or simply lay on my back, doing nothing, thinking nothing, just dreaming into the everchanging clouds.

This was not a refuge for me alone, however. A camp had been built here, and parties from the inn used to rest briefly before pressing on. Reuben and his men had built the huts, and they were marvels of African ingenuity. Native bamboo had been used with loving skill, and neither nails nor screws were used. Even the ropes that held the wickerwork of walls and roof together were twisted of bamboo fiber, made pliable by strong African teeth. The panga had been the only tool in the builders' hands, yet the huts were solid: warm in cold weather, cool in hot, and waterproof, and they blended beautifully into the landscape.

When the camp was first built, the Game Department had contributed a shiny aluminum hut. It was an eyesore and, having been designed for tropical conditions, its excessive ventilation was unsuited to a climate where wind and rain had to be kept out, not invited in. Reuben and his helpers built an armor of bamboo around the monstrosity and topped it with a thatched roof.

From the camp we climbed into the Upper or Hypericum Woodland and right to the top of Gahinga. A great variety of plants thrive here and their fruit, stem, pith, leaves, bark, and roots are important in the gorillas' diet. The yellow-blooming hypericum, from which this area takes its name, is a type of Saint-John's-wort. Here it reaches gigantic proportions and provides gorilla food in abundance, for they eat its bark and roots, even its rotten wood. This plant and hagenia dominate the zone from 9,500 to 11,000 feet.

Above 11,000 feet the trees are stunted from the unchecked forces of the winds. The Upper Forest gradually changes into the Sub-Alpine Zone (11,000 to 12,500 feet). The summit is often shrouded in heavy grey mists, and when the long beards of lichen flutter phantomlike in the piercing winds, the top of Gahinga is no place to linger. This is the eerie region of two bizarre plants, the giant lobelia and giant groundsel.

The lobelia is an extraordinary plant, very different from the garden lobelia. A pipelike stem growing six feet straight up from the ground ends in a cluster of leaves, from which a still stronger stem, as thick as an arm and with steel-blue flowers, rises like a huge candle on an altar. Withered and veiled in mist, these are ghostly plants. They reminded me of toppled telegraph poles in the sandstorms of the Western Desert. One cannot imagine eating that stuff, but gorillas are partial to the bitter marrow of the lobelia stalk. Reuben secretly used to hope that his guests would not heed his warnings about the stalk. He would grin when a visitor would break off a giant candle, take a good bite, then wince as if all the bitterness of the world had filled his mouth. Reuben knew that the bitterness lasted for hours and took away all appetite, and that he and his men would get the sufferer's sandwiches.

The giant groundsel, or senecio, is also one of nature's eccentricities. Several varieties grow on Africa's high mountains, and they are all equally fantastic. The Virunga groundsel has a thicker and softer stalk than the lobelia and is crowned with a tuft of flappy foliage. Its young leaves—green, stiff, and upright—renew themselves from within the top, while the old, brown ones dangle untidily below, like dried tobacco leaves. It has yellow infloresence during the dry season. Only the fleshy base of its leaves is eaten by the gorillas, who arrange the used ones neatly on the ground, as if they had dined on artichokes.

The third giant in this whimsical trio is *Erica arborea*, the tree-heather. It belongs to the same family as European heather, but here it grows into trees well over ten feet tall. There is a veritable erica forest around the brim of Gahinga's crater. On sunny days its shade provdes an ideal picnic spot.

To reach the Alpine Zone (above 12,500 feet) one must climb to the top of Muhavura. At that altitude the lava rock is thinly covered with short grass and moss, a variety of lichen, succulents, and everlasting flowers.

Muhavura's crater, about seventy feet across, is filled with icy water, and once some lunatics in my party went in just to say they had swum at 13,547 feet!

10

Birth of a Volcano

The region dominated by the Virunga Volcanoes, where the gorillas dwell, is one of the most beautiful and exciting sections of Africa. The forces that burst open the crust of the earth created the eight unique mountains, the conical, saucer-shaped hills, and the charming lakes. The eruptions also provided soil of astounding fertility.

These "Places of Fire" stand like a huge wall across the Western Rift. Some geographers maintain that they, rather than the Ruwenzori farther to the north, are the real Mountains of the Moon, which Ptolemy, the Egyptian astronomer and geographer, drew on his famous map of the River Nile in the second century.

Straddling the Rift Valley, the Virunga Chain divides the waters of the two largest rivers in Africa. To the north these waters flow into the Nile basin; to the south they feed the mighty Congo. Visitors were always impressed when I threw a match into the small Kaku River, leaving our Lake Mutanda, and told them of the long journey this tiny piece of wood was setting out upon. The Kaku River, I explained, would soon join the Rutshuru River, which, in turn, would take the match into Lake Edward. It would then float down the Semliki into Lake Albert, leave by the White Nile at Pakwak, travel through the Sudan to Khartoum, then down the Nile proper via Abu Simbel, Aswan, Luxor, and Cairo until it finally reached the Mediterranean in the delta somewhere near Alexandria.

Half a million years ago a tremendous volcanic convulsion forced up Mt. Mikeno and Mt. Sabinyo, the two oldest mountains of the chain. A short time as volcanoes go, the 500,000 years have gnawed and eroded the slopes of the mountains so that today both are craterless. *Mikeno* means the "bare one" and its spectacular rock summit (14,500 feet) is merely the central cone of the original huge crater. Of Sabinyo (12,000 feet) five corroded teeth are all that is left of the former crater mouth. The natives call the mountain "Father of Teeth."

A long rest followed the creation of these two giants. Then, about 100,000 years ago, all hell broke loose again. Enormous masses of lava built up four new cones, which still have craters and look like volcanoes. Mt. Mahavura (13,500 feet), the "Guide," lies to the east; it is seen from afar and seems to point the way. Next to it is Mt. Gahinga (11,400 feet), its name meaning "A Heap of Stones." In the center, with the lesser Vishoke in front, stands Mt. Karisimbi (14,780 feet), the "White Shell." When this peak is sprinkled with snow, the white ridges between its numerous canyons radiate like the ribs of a cowrie shell.

After a second, shorter lull, the eruptive forces came to life again about 20,000 years ago and created Mt. Nyamulagira (10,000 feet) and Mt. Nyiragongo (11,000 feet). These are the only volcanoes of the group that are still active. Nyamulagira's crater is about 500 feet deep and more than a mile across. It has an outlet on the southern side through which, during most eruptions, the lava flows down toward Lake Kivu.

The crater of Nyiragongo, two-thirds of a mile in diameter and 1,200 feet deep, is shaped like a kettle, with perpendicular walls dropping down 400 feet to a shelf.

The daring Earl Denman, who climbed all eight volcanoes barefoot, descended alone to Nyiragongo's shelf, using two ropes tied together. He staked his life on the tenacity with which these ropes were held by the inexperienced Africans standing on the crater's rim. The double rope proved to be too short, but Denman continued even without this security. However, he was soon driven back by a panicky fear that a change of wind would blow the sulphurous vapors near, and poison him.

The men holding the rope were relieved when Denman's head appeared over the edge of the cauldron, for, to their thinking, it was sheer lunacy to tease the spirits of the "Places of Fire." The African inhabitants of this region have a congenital fear of anything resembling hot lava or smoke.

Denman himself must have tasted that same fear as he stood alone on the first shelf and did not venture lower. From that vantage point he could see that the kettle was not as empty as it had appeared to be from the top. On the bottom, 800 feet below him, a bubbling lake of black lava with two chimneys steamed with alarming force. It was a terrifying sight.

The American zoologist, George Schaller, spent a night on Nyiragongo and wrote of his feelings of awe:

A soft red glow suffused the sky as I crawled on all fours to the crater rim. I lay there and stared at a sight so beautiful that my heart wanted to cry out and waves of shivers moved over my back. Far below, the lake of lava shone a bright orange-red, and as the glow rose and spread outward, it turned a delicate purple. I was an intruder, spying under the cover of darkness into a purple vortex at the center of the earth lying before me naked, its emotion laid bare.*

Major eruptions occurred in 1938 and 1948. Old-timers, believing in a cycle of three, prophesied a third big blow-up for 1958. It came—and I was there to see it!

*George B. Schaller, *The Year of the Gorilla* (Chicago: University of Chicago Press, 1964).

Nyamulagira began to show signs of activity in 1958, but, strictly speaking, it was not the volcano itself that was cooking. A new satellite was being born in its region and unmistakable labor pains heralded the phenomenal advent. For some months before the actual birth, violent tremors reproduced the sensations of sailing in rough weather. Tables and chairs slid suddenly across the floor. Drinks were spilled on the bar counter. Sleep was disturbed in the night by uncanny forces rocking the beds. The tremors were accompanied by a sound like the roar of an approaching subway train. It was a sinister sensation, this paroxysm in the bowels of the earth as the tremendous pressure underneath tested its crust for a convenient outlet.

It is said that gorillas, elephants, and buffaloes sense approaching volcanic disturbances and leave the danger zone long before any such symptoms are detected by men. This may be true, but I am doubtful. Gorillas, for instance, have no need of divining powers. They simply seem to shun volcanic territory, although climate and vegetation are much the same there as in the region of the six extinct volcanoes. Perhaps an ancient subconscious memory of former catastrophes, an inherent wisdom, guides them.

As for elephants, I give them the benefit of the doubt. The warnings that preceded the birth of Kitsimbanyi were strongly felt by humans. Why not also by the elephants, who live so much closer to nature? Should not the animals, therefore, sense the slightest motion of the ground, scent the least change in the air, when even our dulled senses perceive these forebodings of disaster? Weeks before this eruption many elephants were seen deliberately crossing the Rutshutu-Goma road at a place where, at other times, only an occasional solitary bull was seen. The pachyderms obviously were changing over from the threatened side of the road to the safety of the extinct volcanoes on the other side. This sudden exodus may have been just a coincidence. Reliable reports testify that on some previous occasions the elephants were taken by surprise, cut off from their retreat by advancing lava streams and doomed to die a gruesome death. Smaller mammals, reptiles, birds, and insects do not try to escape in time, and many thousands are burned alive.

Dr. Jacques Verschuren, the Belgian biologist, knows the Albert Park and its wonders better than any other living person. He camped there hundreds of times and penetrated places where no one before him had ever been. Once he came to a spot that looked like the legendary cemetery of the elephants. He was flabbergasted at first by an accumulation of animal bones, remains of large and small species, but soon discovered the cause of the mass destruction. Poisonous fumes were constantly seeping from cracks in the earth. Any living creature that approached too closely was asphyxiated. He himself withdrew with all possible speed.

During this century, Nyamulagira and Nyragongo, only ten miles apart from crater to crater, have produced enormous quantities of lava, as if in brotherly cooperation. It seems as though they are determined to fill Lake Kivu, for all their eruptions have sent torrents of burning lava down to the lake with devastating speed. Six to eight miles long and half a mile wide, those burning flows sometimes heated the water of the bay of Saké to such a temperature that, for months afterward, fishermen found the fish floating on the surface, already boiled.

The little port of Saké, cut off now from the lake and no longer accessible to the steamer, is surrounded by a fantastic landscape. The dead streams of lava lie like twisted coils, like a solidified discharge from a fatal wound in the tortured earth, ugly black scars marring the beauty of the once pleasant bay.

But nature recovers quickly. A few blades of grass, some lichen, mosses and ferns, and a bit of brush soon sprouted from cracks in the hardened crust. Some of the older flows, in a state of decomposition, are already covered with a thin layer of green. The history of the volcanic activity of the two mountains can be read clearly from the various vintages of lava and from the vegetation that has sprung to life on these desolate black rivers from the underworld.

On their way to the lake these lava flows cross the main road from Goma south to Bukavu and block the traffic for months. The crust soon cools and hardens, but underneath the lava remains soft and hot. In dry weather one can pass over safely on foot. When it rains, however, the rain water penetrates the porous lava and deadly vapors rise from the fissures. Despite warnings, there are always some natives who attempt to cross the flows during or shortly after a rain. Their bodies testify to the deadliness of the fumes. The main craters, although they themselves have not erupted for ages, are by no means dead. Their chimneys, leading down into the "lava factory" for thousands of feet, are not choked forever like those of Muhavura, Gahinga, and Visoke. They still produce lava prodigiously. On many occasions their craters, particularly that of Nyiragongo, have been full to the brim. Instead of overflowing, however, they have discharged their vast volumes through other vents, as their inherent energy blasts open new passages on their flanks or at their feet. A secondary crater or an entirely new satellite is thus created. This is the origin of the many cinder cones or miniature volcanoes. Now terraced, these hills give the landscape its peculiar moonlike aspect. The life of these cinder cone volcanoes is usually short. During the few months of boisterous activity, they build up to a height of some 300 to 400 feet, then they die forever. Kitsimbanyi, the "unexpected," was just such a satellite, and I witnessed its birth and death in 1958.

The actual birth of Kitsimbanyi was observed by two tribesmen. Warned by the vehement tremors and rumblings in their vicinity, the two men had been on their guard. They noticed suspicious fumes rising from a patch of bush, which was far away from the "Big Smoke," as they called it, and in an area where they had never before seen any fire. As they approached the smoking bushes, a wave of scorching heat drove them back. The ground under them began to tremble and to split open at their feet. At first, many small cracks appeared on the surface. Then these fissures opened wider. Suddenly, there was a sharp hiss and a huge chasm opened a few yards away. From its depths smoke and flames shot up. The bush around was soon ablaze and a thick porridge came oozing over the edge of the hole. It looked, they said, like *posho* overflowing a cooking pot. The whole incident lasted only a few minutes. The men who saw it ran away in terror.

Although the eruption which created Kitsimbanyi in 1958 was less impetuous than the eruptions of 1938 and 1948, the performance caused great excitement in the neighborhood. Many sightseers poured into Goma, the starting point for safaris to the new volcano. Long convoys of tourists were driven close to the crater where the antics of young Kitsimbanyi could be observed conveniently.

I doubt, however, that the same privileges were extended to non-Belgian scientists, for on previous occasions foreign geologists had been refused permission to approach the scene of the eruption. Renowned specialists from Entebbe, South Africa, and Switzerland had to study the phenomenon miles away, while Belgians could view the grand spectacle stage-front. The parochial Belgians, it seemed, wanted to cover the eruptions by themselves.

I shared the disappointment of the experts, for I, too, felt frustrated. However, I could not blame the Belgians for the arm I had broken shortly before, while exploring a cave. The limb was stretched upright in a cast and no normal car could accommodate me. I had to content myself with viewing the birth of Kitsimbanyi from a hill behind the inn. Every night we climbed up to watch. Even from a distance of some thirty miles, the flaming, furious volcano was an impressive and profoundly disturbing sight.

As if blown by bellows in a forge down in the earth, a tremendous cone of fire belched up from the crater into the black night, then dissolved into a fiery splatter. Projectiles of glowing red rocks were flung up and smaller ones shot rocketlike into the sky, burst, and rained down in a myriad of yellow and red balls of fire. Then came a short lull, a deep breath, strangely human. Again a tremendous fountain of fire, jetting up a few thousand feet, sprayed sparks for

The dead Saza Chief being carried down from the mountain. Courtesy of Ian McClellan.

An adult female, Effie, of Group 5, on a trail near Dian Fossey's camp. She glances back as the group feeds slowly away from the observers. Courtesy of Jay H. Matternes.

The dominant silverback (male), a magnificent animal called Uncle Bert, of Dian Fossey's Group 4, is sitting on a trail among stinging nettles eating the ubiquitous vine galium. Although the photographer was as close as fifteen feet from Uncle Bert, the ape displays only mild interest as long as he knows where the observer is—a tribute to the excellent work of Dian Fossey in habituating these animals. Courtesy of Jay H. Matternes.

miles around and set the forest alight. It was a fantastic pyrotechnical display, expertly produced by an angry giant in Hades.

We also could clearly see the lava stream wriggling down the slope like a luminous serpent, leaving desolation and destruction in its wake. Gushing over the crater's edge in yellowish-white cascades, the torrent changed to a deep red at a lower altitude. This lava stream, instead of aiming south toward Lake Kivu as expected, turned east toward us, threatening to cut into the Rutshuru River and change its course.

After five lively months young Kitsimbanyi died. It had reached a height of 250 feet without causing serious damage. The lava stream had spared the Rutshuru River and no coffee plantations were destroyed. It had been a breathtaking sight, the birth of a volcano, and a disconcerting experience to watch nature at work, to see her unharnessed energies destroying, changing, and reforming the face of the earth.

A view of the picturesque Kisoro valley, Uganda, with its thatched-roof huts. The trees are nonindigenous eucalyptus. Courtesy of Jay Matternes.

Walter Baumgartel and Mutwale Paulo, the Saza Chief. Courtesy of National Geographic Society, Washington.

Walter Baumgartel's inn, Travellers Rest in Kisoro, as it was when it served as the informal headquarters for gorilla talk in Uganda. Courtesy of the author.

Reuben Rwanzagire, the expert gorilla guide and tracker. Courtesy of the author.

An old face of Africa, a gray-haired man from the Bantu tribe. Courtesy of
Gerhard Gronefeld, Munich.

The cauldron of Mt. Visoke, an inactive volcano of the Virunga chain. The mountain is 12,172 feet high. The trees in the foreground are giant senacios, whose exceedingly brittle branches are occasionally eaten by gorillas when they range to the higher altitudes. Courtesy of Jay Matternes.

The author with a skull and a death mask of a gorilla.

(LEFT)
Gorillas have an astonishing range of shapes and sizes. They may be thin or fat, and their facial expressions vary greatly. Despite his beard and wrinkles, this is a young gorilla, about four years old. Courtesy of National Geographical Society, Washington.

Little Reuben, who was found in the gorilla forest at the side of his dead father. He was later sent to Regents Park Zoo, in London.

Goma, born in the Basel Zoo and raised in the house of Dr. E. M. Lang, the director of the zoo. Courtesy of Fee Schlapper, Baden-Baden.

Gorillas of Group 4 near Dian Fossey's camp during a midmorning siesta. An old female, Flossie (left), looks somewhat apprehensively at an observer who has just come into the clearing. Courtesy of Jay Matternes.

The two romping juveniles, Tiger and Augustus, are also a part of Dian Fossey's Group 4. Courtesy of Jay Matternes.

(RIGHT)

Pucker Puss, a captive two-year-old female gorilla, and CoCo, a sixteen-month-old male, enjoy a frolic with Dian Fossey, the young American woman who is now studying mountain gorillas in Rwanda, Africa. Courtesy of the National Geographic Society, Washington.

72

11

Gorilla Tracking

Gorilla tracking was strenuous business, requiring good legs and lungs and a sound heart. But when the search was crowned with success, the thrill of achievement more than offset the effort.

No one could accuse us of exploiting our protégés. On the contrary, some people thought we were too cautious. We refused to take more than two or three persons at a time and would make only one tracking safari a day. We tried to avoid upsetting the apes, and I am sure our discretion helped keep them from fleeing down into the forests on the Rwanda side.

Jill's theory held that it was wrong psychologically to track these touchy animals by stealth, avoiding all noise. One should approach them openly and naturally, she insisted. The animals would then feel less suspicious.

Reuben took a different approach. His own advice was, "Don't wash yourself so often. The smell of your soap and toothpaste makes the gorillas nervous and drives them away." Perhaps both Jill and Reuben were partly right!

The gorilla depends mostly on his sight. His senses of smell and hearing are not as acute as those of many other animals, although they are probably superior to ours. I have no doubt that the gorillas could hear and smell us long before we could hear and smell them.

They certainly could have quickly detected large parties of visitors and silently melted away. It is difficult to control a crowd of people; I was always afraid a straggler might get between a male and one of his wives, or between a mother and her child. If that had happened, an alarming situation would ensue. Peter, one of the trackers, knew this from experience. He had once stepped on the toes of a female gorilla he had not noticed in the thick vegetation. She screamed in terror. Peter, not usually a nervous person, also screamed and bolted down the slope as fast as he could. Luckily, the silverback leader was some distance away. By the time he arrived at the scene, Peter had landed safely in a thistle bush.

Reuben and his men usually knew where a particular family or group was living. Although gorillas are nomads and sleep every night in a different nest, they often stay for weeks in the same general area. Then suddenly, for no known reason, they are off and gone. Other nomads, such as elephants, wander on after a feeding place has been eaten bare. Gorillas often leave theirs when food is still plentiful.

Sometimes they climbed far up Muhavura to an altitude that offered only a frugal diet of lichen and mosses, which hardly could be tempting to such fussy eaters. Why? Was it mere wanderlust? Or do these plants contain some nutrients their bodies need? Sometimes, for equally inexplicable reasons, they trekked down into Rwanda, often returning, just as erratically, the following day.

A gorilla search, therefore, was often difficult and required much skill and patience on the part of the trackers. To save time and energy, Reuben would send his trackers out at daybreak to locate, as a starting point, a group of nests where a gorilla family had spent the previous night. Gorillas are not early risers. They were often still asleep when the trackers found them, and the men were careful not to disturb their slumber. The trackers would leave marks along the way for Reuben to follow and he always managed to find the men, even in the thickest forest. The trackers then led the party, as directly as the terrain allowed, to the nests.

From there it was easy to follow the morning's fresh trail. Gorillas don't bother to disguise their spoor as many other animals do. Their droppings are abundant proof of an enviable digestion and they leave remnants of their meals in neat piles behind, convenient milestones to follow.

Countless gorilla, elephant, and buffalo paths crisscrossed our mountainside, and the sureness with which Reuben and his men picked the right tracks from this maze always astounded me. Reuben would bend down to study a footprint or knuckle mark, or to examine droppings for color and temperature to determine how long before the animals had passed. A trodden blade of grass, a trampled plant or a nibbled piece of bark would show him the direction taken. On hard ground the spoor was sometimes completely lost—or ran in confusing circles. In such instances Reuben's skill amounted almost to wizardry. He sent his trackers out in different directions. Soon birdlike whistlings would be heard and Reuben would indicate that the right trail had been found. The natives' eyes were sharp, their minds observant. Not the faintest clue escaped their notice. If they disagreed among themselves, all possibilities were logically discussed, but Reuben was invariably right. I always marveled at his uncanny native craft, but I never mastered it.

If the gorillas happened to notice us early enough they would disappear noiselessly. Usually, however, the leading male detected the intruders too late for such a safe retreat. Then he would send his family ahead, while he, the protector, remained behind, waiting in ambush to give these unwelcome intruders a thorough fright.

This game of hide-and-seek often could be exasperating. We would get worn out and thoroughly fed up. Then Reuben would suddenly sniff the air—and declare that he smelled gorillas. Our spirits would revive, although our own less sensitive noses detected nothing but the strong scent of herbs. As we moved on, we gradually would become conscious of the musty, pungent, yet sweetish and not unpleasant odor peculiar to gorillas. We knew they were close and we always felt excited and a little afraid.

I had become familiar with the bluffing tricks of male gorillas. The tactics the apes employed were always similar. The father or leader would allow us to come close, then try to check any further advance by frightening us. Sometimes he varied his scare techniques and actually would jump at us. This was only a mock attack, but was most disconcerting. Even Reuben and the trackers would feel uneasy and tense. Peaceloving though he is, a gorilla, finding himself genuinely cornered, will sometimes attack. You cannot trust such agitated animals.

Even less, however, could I trust members of my own party. We know gorillas are moody and unpredictable, but I doubt if they ever can be as foolish as humans. Our guides had to be prepared for any unpredictable human antics.

One eccentric guest was seized by a fit of uncontrollable laughter when the great silverback, Saza Chief, roared at him. The Chief was obviously in a savage mood that day and beat his chest like a maniac. Only a few steps separated us from the massive gorilla. I did not appreciate the volley of hysterical laughter that greeted every drumroll. The Chief looked perplexed, even disgusted, as he withdrew as usual into the undergrowth. What dignity! I would not have blamed him for reacting violently to the undeserved ridicule.

Another guest, on the eve of his safari, heard an improbable yarn about a man who threw a tuft of grass at a gorilla he met in the Kayonsa Forest, whereupon the gorilla threw the tuft back at him and a fine game of ball ensued between man and ape.

As our party crawled toward a barking male the next morning, I noticed that this guest was holding something grassy in his hand. As we came close to our quarry he stood up, and only then did I realize what he proposed to do. I stopped him just in time before he threw his tuft to Saza Chief.

Horst, my German actor friend, fell madly in love with Africa. He liked nothing better at Travellers Rest than to be sent on an errand into the Gorilla Sanctuary and spent much time there, supervising the reconstruction of the saddle camp. He often accompanied visitors and got to know the gorillas well. A gifted impersonator, he soon was able to add a lifelike imitation of Saza Chief to his repertoire, which already included Hitler and a number of famous German actors. My friend was particularly interested in breath control and voice projection and was convinced that actors and orators could learn a great deal from the gorilla's astounding techniques.

Horst was a guest on one unforgettable excursion in which the leading male gorilla outflanked us and reappeared unexpectedly behind our backs as we stood in a narrow ravine. Even Reuben and the trackers didn't like the looks of the angry ape towering above us. Seldom had I encountered such rage before: He barked, roared, screamed, and screeched; trampled, drummed, and shattered branches. Then—he leapt toward us! At that moment I wished I had never come to Africa.

With horror I realized that Horst was about to put on his own act. He rose slowly, thrust up his arms, cupped his hands, and was ready to deliver the first drumroll on his chest before I managed to pull him down and quell his first gorilla scream by putting my hand firmly over his mouth.

Heaven knows what would have happened to Herr Horst had I not stopped him from demonstrating his talent. The gorilla probably would have thought he was a challenging male and the German press would have been able to splash the following on the front pages:

PROMISING YOUNG ACTOR CRUELLY KILLED
BY ANTI-GERMAN GORILLA IN UGANDA

I often thanked the spirits of the mountains that my gorilla-feeding experiments had ended in failure, for these safaris were much more valuable and interesting.

One day the representative of a worldwide insurance company arrived at Travellers Rest and offered to put at my disposal any reasonable amount of capital to make the inn a first-rate hotel, transforming Kisoro into an important tourist center. The conditions he stipulated were that a "treetops hotel"—similar to Kenya's famous Treetops at Nyeri—be erected up in the Sanctuary, and that a cableway be constructed to make the ascent possible for any elderly millionaires who wished to see the gorillas.

I'm proud to say that I told this tempter that the Game Warden would have locked me up in a lunatic asylum were I ever to suggest such a scheme. I would rather have eaten *posho* and beans like the natives and walked about in a goatskin than turned the Gorilla Sanctuary into a peep show.

12

Hunters
and
Collectors

Gorillas were known as far back as the fifth century B.C. Hanno, a Carthaginian admiral, sailed along the west coast of Africa and reported that when he and his crew went ashore, presumably at what today is Sierra Leone, they met a tribe of native savages, consisting mainly of hairy women. His sex-starved sailors tried to capture some, but the "ladies" resisted fiercely. They used their teeth and nails so furiously that the attackers could only escape their wrath by clubbing them to death.

Our mountain gorilla was not discovered until 1902, when a Captain von Beringe shot the first specimen on the slopes of Mt. Sabinyo, probably on the Rwanda side, which was then part of German East Africa. The Captain and some friends had gone to climb the hitherto unconquered Mt. Sabinyo and, having no rope with them, had been forced to give up the attempt. As they camped on a narrow ridge at 9,300 feet, they watched a troop of large black apes above them succeed in reaching the top. Von Beringe shot two but could retrieve only one of the bodies from the deep ravine into which they had fallen. About five feet tall and weighing 220 pounds, this specimen was a young male of moderate dimensions, but he was larger than any apes the German had ever seen. Von

Beringe sent the skin and skeleton to Professor Matchie, an expert in the art of classification at the University of Berlin, who recognized the ape as a gorilla.

This discovery caused a sensation. A few years earlier, when the okapi had been discovered in the forests of the eastern Congo, scientists had believed that the Black Continent had nothing new to reveal, or at least no more large mammals. They were convinced that gorillas could live only in the hothouse climate of the western forests. This slightly heavier, jet-black fellow, living in the high, cold, inhospitable mountains more than a thousand miles away from his West African kin, threw their whole theory out of gear.

Professor Matchie, who saw in the slightest deviation from the norm a new species, promptly named the new ape, in honor of its discoverer and of himself, *Gorilla gorilla beringei matchie*. His name was removed later when an American, Harold Coolidge, working on the taxonomy of the gorilla, established that there was only *one* species, split into two subspecies, the lowland or western gorilla, and the mountain or eastern gorilla. The latter has recently been subdivided again into two groups, the mountain gorilla and the eastern lowland gorilla. The differences that exist are due mainly to local conditions and environmental influences. All three subspecies can mate with one another and are difficult to identify accurately if only one species is available.

Speke reported that, when moving north in 1861 in quest of the sources of the Nile, his porters told him of wild, humanlike monsters living in the strange mountains on the western horizon. Perhaps it was a happy accident that Speke himself was so obsessed with solving the riddle of the Nile. Otherwise, mountain gorillas might have been discovered forty years earlier—and not lived long enough for us to know them.

In the 1920s the mountain gorilla suddenly drew a host of hunters and naturalists to the eastern Congo. They came to collect specimens by catching or killing the animals, and many gorillas lost their lives. Some useful observations were made, but they were incidental and in no proportion to the sacrifice. Perhaps one should not judge these men too severely. Gorillas were big game in those days; wildlife protection was still in its infancy and few people had any scruples about hunting the rare apes. Prince Wilhelm of Sweden, for example, collected no less than fourteen of my "cousins" on his famous expedition of 1920-1921 to the Virunga Mountains, as reported in *Among Pygmies and Gorillas.*

Many years later, an elderly guest at Travellers Rest was discussing mountain gorillas with me. He was still spry and youthful, and his intelligent questions convinced me that he knew what he was talking about.

"You seem to know something about gorillas," I said.

"Why shouldn't I, after shooting fourteen of them!" he exclaimed.

"Fourteen gorillas!" My heart sank.

"Well, it was forty years ago and on the other side of the mountains in the Congo," he said, a little sheepishly. "I am Kenneth Carr, and I accompanied Prince Wilhelm of Sweden on his expedition. I am ashamed of it now, but then it was a great adventure. I was young, and I loved to hunt. The Prince paid well and I really thought nothing of killing the apes."

The Prince and his party, Carr said, actually camped where I later built my huts, on the saddle between Muhavura and Gahinga. There had been groups of gorillas in the vicinity, "but they were much too clever for us," he said. They had always managed to escape in the devilish undergrowth.

The noted American naturalist, sculptor, and taxidermist, Carl Akeley, "collected" five gorillas in the same area in 1920-1921. He was, however, a chip off a different block. I have always admired him, even forgiven him for his gorilla-hunting expedition. He grew very fond of these apes and stopped killing after he had collected his five specimens. He suggested to King Albert of Belgium that the Karisimbi-Mikeno area should be declared a gorilla sanctuary, open only to scientists studying the life of these apes in the wild.

Carl Akeley loved Africa and made it his life's task to preserve the wonderful world of its wild animals for future generations. His dream of an African Hall in the American Museum of Natural History in New York City was fulfilled. Lifelike groups of all large African mammals can be seen there in authentic African settings. Such habitat groups were something entirely new in Akeley's time. He was a perfectionist and insisted that every detail be accurate. He himself collected shrubs, grasses, plants, twigs, and stones and required that the artist who painted the backgrounds make his studies on the spot. Fourteen breathtaking scenes of African wildlife are now on exhibit in the Carl Akeley Memorial Hall of the museum, with its spectacular elephant herd dominating the main floor. There resides the impressive group of the five mountain gorillas—three males, one female and one infant—which he had collected on his fourth African expedition in 1921.

On that expedition he also succeeded in shooting several hundred feet of film of live gorillas in their native forests, the first ever taken of the rare apes.

Five years later he returned to Africa, accompanied by his wife and William Leigh, the artist who was to paint the appropriate backgrounds, the magnificent panorama of the volcanoes as seen from the spot where the huge chest-drumming male of the group had been shot on the previous expedition.

Akeley's joy at being back in this African fairyland did not last long. This time he had come to observe and study gorilla behavior, but no sooner had he arrived at the final camp at 11,600 feet when he fell ill and died a few days later of fever and fatigue. However, it was a perfect end to an explorer's life. As he had wished, he was buried in the meadow below his final camp and a simple marker, Carl Akeley, November 17, 1926, was placed there.

Akeley was a hunter and collector rather than a scientifically trained behaviorist. His observations, however, added to the limited knowledge of the mountain gorilla and he did much to destroy the notion of their ferocity. His suggestion for a gorilla sanctuary was carried out by the Belgians. A few years later the British followed suit and declared the Uganda section of the volcanoes a gorilla sanctuary also. Without Akeley's legacy, the sanctuary, the threatened gorilla species probably would have disappeared from these mountains long ago.

13

Unofficial Gorilla Headquarters

Guiding tourists through gorillaland was a thrilling enterprise. I truly believe that it did, in a small way, help make the great apes better understood. I soon realized, however, that only long-term, methodical observation by trained students would lead to answers to basic questions about gorilla behavior and ecology. These two branches of biology had emerged only recently and comparatively little investigation had been done in their fields.

Before I left Travellers Rest, experts in behaviorism and ecology were flocking to Kisoro, some of them intent upon gathering knowledge about species very different from the gorilla. There was a weaverbird specialist who busily visited all countries where these feathered architects built their astounding nests. I remember a zoologist who concentrated on grasshoppers; and I could not help admiring the zeal of the frogmen who sacrificed their dinner every evening, for it was served at precisely the hour when the objects of their investigation woke up and started their nightly concert in the swamps along the lake shore. I was interested in all these scientists, but my goal was to attract those who sought the truth about our gorillas.

Jill was a competent observer of gorillas but she wanted only to complete the year of fieldwork that Rosalie had begun. I also realized that no single observer could make all of the necessary studies. Therefore in 1956 I approached a few scientific institutions suggesting that they send a team of experienced observers with a special interest in primates. The species was threatened with extinction, I warned. No time should be lost in making the study. I pointed out that no long and costly safari was required. The scientists could reach the gorilla habitat and begin their observations on the day of their arrival. My thought was that three people working perhaps in three teams with two native trackers each, and augmented now and then by a botanist or other specialist, would suffice. At little cost, they could be made comfortable in our camp on the saddle. They would be just a few hours from the Kisoro base and so could easily replenish their supplies.

Sir Solly Zuckerman, the prominent British scientist, responded to the appeal I addressed to the London Zoological Society. He promised to raise funds for such an enterprise and asked me to wait for further information from him before approaching anyone else. Months passed, however, and I heard nothing more.

Meanwhile, expenses mounted for car, camp maintenance, and wages of guides and trackers. When my resources ran out I ventured to approach Sir Solly again, asking if he could send some interim funds to bridge the time until his expedition was ready. To my chagrin, I discovered that he had soured on the scheme. Scientific observation, he wrote, had to probe more deeply than we had done. Our amateurish effort did not justify financial support.

I knew that myself. It was for that very reason that I was trying to persuade scientists to come and raise to a level of expertise an enterprise that had started as a mere tourist attraction and had grown too big for me to handle.

At that time I was still out of action with a broken leg. It was impossible for Jill alone to follow the ever-moving groups of gorillas over the vast and difficult terrain of the mountains and observe their social organization and sexual behavior.

I knew that the gorillas' sexual behavior was a field in which Sir Solly, the reluctant British scientist, professed to be particularly interested. He had studied the sexual behavior of primates in the safety of the zoo, a fairly simple matter. I was amazed that he did not jump at the opportunity to check his conclusions by observation and study of creatures living freely in their own domain.

I was convinced that Rosalie and Jill, amateurs though they were, had opened up fields any scientist should be eager to explore. Indeed, Rosalie actually had seen a gorilla couple in the process of copulation. No iron bars had stood between her and them. Apart from myself, no one, to our knowledge, had observed such a scene in the wild before. Rosalie had, in fact, missed the significance of her discovery at the time. She had thought the two were just playing, so she had not been unduly worried about her own safety. Only later, when she described the scene to me, did it suddenly dawn on her that she had probably witnessed the holy act of mating. And as Reuben described the scene, sharp observer and gifted actor that he was, no doubt could remain as to what it had all been about.

Left in the lurch by Sir Solly, I turned in desperation to Professor Raymond Dart in Johannesburg, whom I knew to be a man of vision and enthusiasm. He was engaged in a bitter controversy with the anthropological elite of the time about his australopithecus, the oldest humanlike fossil then known. In a fossil skull brought to him in 1924 Dart recognized an early ancestor of ours who had lived in southern Africa between one and two million years ago. Dart propounded that the creature had walked erect, had been a hunter, eating meat and making weapons, whereas the experts declared it had been nothing but a small ape, probably a chimp. However, amazing discoveries made in recent years in Tanzania, Kenya, and Ethiopia by the Leakey family and other expert fossil hunters have substantiated Dart's claims and his australopithecus is now generally accepted as a hominoid who opened up entirely new vistas into man's origin and evolution. Professor Dart immediately recognized the scientific possibilities of our Kisoro effort and the boon it would be to work in a place where everything was ready for the study of the rare and inaccessible great apes. Although his own university could not send an expedition at the time, Dr. Dart started his Gorilla Research Unit, which provided a regular monthly grant for our efforts.

As long as the work was being done and science served, Professor Dart did not care who received the credit for the enterprise that he had made possible. As long as there were enough gorillas in the Sanctuary for the scientists to study, Dart and his successor, Professor Philip V. Tobias, a prominent anatomist, kept our work alive with their steady financial support. Their friendship and moral support were equally encouraging.

With financial aid assured, I approached Sir Solly again and suggested combining forces, for I felt that he could profit from cooperation with Dart. He did not even reply.

When Dart visited Kisoro in December 1957, I mentioned my correspondence with Sir Solly. He roared with laughter: "That is priceless, Sir Solly benefiting from me!" I hadn't realized that I had stepped into a hornet's nest. Sir Solly was a bitter opponent of Dart's anthropological theories.

During Dart's visit, we took him up to see the gorillas, and Saza Chief, as if conscious of the honor, delighted our benefactor by a splendid show, roaring and drumming in his best form. Although well over sixty at the time, the professor climbed and crawled about as nimbly as a youth. He tried all the plants on which gorillas feed; he even tasted the bitter pith of the lobelia and other exotic roots and vegetables, and asked us to take down a complete collection of those greens, to be prepared as vegetables or salads at each meal.

Professor Dart seemed to approve of the stuff, but the other guests were less enthusaistic. Only the tender bamboo shoots were an all-round success. After that discovery, I no longer had to worry about what to serve yogis, Adventists, and other vegetarians during the dry season, when no fresh vegetables were available. Bamboo shoots served with hollandaise sauce or cooked in chicken broth became a regular specialty on our menu. Many guests preferred them to tinned white asparagus.

Sir Julian Huxley, the eminent biologist, and Lady Huxley were our guests in September, 1960. I was honored to have them visit Travellers Rest. Because of the work of his grandfather, the great Darwinian Thomas Henry Huxley, and his own fundamental research on genetics, Sir Julian had a deep interest in evolution, in the kinship of men and apes. They told me that they had been in the Virunga Mountains in 1929, long before there was a motor road into the Eastern Congo. To visit Carl Akeley's grave, they had had to walk from Kabale, via Rutshuru, to Kabara. They had hoped to see some of the gorillas on that trip but luck had been against them. Now, after so many years, they had come to Kisoro with high hopes of meeting our apes. Reuben had seen some near our camp the day before the Huxleys' arrival, and chances of fulfilling our guests' wish seemed auspicious.

The next morning our party puffed up to the saddle only to find that the gorillas had moved higher up. The steep slope with dense vegetation was too difficult for the elderly couple to climb, although Sir Julian did extremely well for a septuagenarian. When rain threatened, they decided to give up. They returned to the inn a little disappointed, of course. But they had seen gorilla nests and droppings, they had tasted gorilla food plants, and they had deeply enjoyed being back in that "amazing country."

Another visitor whose major interest was the origin of man was Robert Ardrey. He came with an introduction from Professor Dart, and he was accompanied by Berdine, his South African wife and illustrator of his books. Dart had told me that the couple wanted to "talk gorilla," so of course they were welcome.

Ardrey had started his career as a dramatist. *Thunder Rock*, a fine play written some twenty years earlier, had made a deep impression on audiences in many parts of the world. I was puzzled that Dart should send me a playwright. Why did Ardrey want information on gorilla behavior? Did he intend to write a play about them? I soon discovered that he had abandoned playwriting and had become a serious student of evolution.

As a dramatist, Ardrey had depicted human behavior on stage in great city theaters. The stages on which his new dramas were to take place were the remote caves of the South African high veldt. His theme was to be the much more complex and exciting story of human origin and evolution. There, at the sites of the very beginnings of man, he perceived the first manifestations of human nature and behavior.

Ardrey began to acquire the scientific tools he needed for competent investigation. He went from library to library, museum to museum. He visited the game parks in East Africa to observe animal behavior. He began to write his fascinating books.

On his visit to Travellers Rest Ardrey had with him the manuscript of *African Genesis: A Personal Investigation into the Animal Origins and Nature of Man.* That disturbing book was published in 1961 and it has caused much controversy. It was followed by *The Territorial Imperative* (1966), *Social Contract* (1970), and *The Hunting Hypothesis* (1976).

Backed by the grant from South Africa, Travellers Rest soon became the Unofficial Gorilla Headquarters in Africa, as George Schaller later called it. Scientists could come to Kisoro easily, find a comfortable home base, and observe gorillas living within easy reach.

The first scientist to avail himself of our facilities for the purpose of actually studying the habits of the great apes was Dr. Niels Bolwig, a lecturer of Zoology at Witwatersrand University in Johannesburg. This likeable Dane visited us early in 1959. He was interested chiefly in the nest-building methods of gorillas and made a thorough study of the subject. He even did some nest-building himself, first in our mountains, then in the impenetrable forest. He later wrote a study in which he stated that the clockwise—or was it counterclockwise?—direction in which the bamboo canes were bent proved that our gorillas were left-handed, whereas those in the impenetrable forest were right-handed. However, before he had ended his dissertation—if I understood it correctly—he had disproved his own theory: All the apes were right-handed, it seemed; only some had a definite tendency toward the left!

While he was with us, Niels conducted his researches thoroughly—but at times with a certain nonchalance. One day, in the sanctuary, Reuben left the doctor alone while he went off in search of more nests. The doctor was deeply absorbed in the study of a group of beds. When Reuben returned, the place appeared deserted. Where was the doctor? Reuben, already worried, heard a strange noise above his head. Could it be a leopard growling? He was relieved to discover the nest-building scientist happily ynoring up in a bamboo bed. Bolwig had climbed up to test the nest for strength and found it as comfortable as a spring mattress. It was so inviting that he had fallen fast asleep.

Our efforts in Kisoro drew the attention of specialists in the field throughout the world. The New York Zoological Society sent Dr. Harold S. Coolidge of the American Academy of Sciences in Washington to investigate the possibilities for a thorough study.

Coolidge was familiar with the country. He had collected a male gorilla in 1927 in the Eastern Congo for the Museum of Comparative Zoology at Harvard. Later he had worked on the taxonomy of the mountain gorilla and was able to reduce the many species to only one, which he subdivided into two races, the Western and Eastern ones.

He spent only one day at Kisoro, saw an enormous male gorilla at close range and said he would recommend our place as one suitable for studying gorilla behavior. The African Primate

Expedition—or APE—was the result of his preliminary work. It was well prepared and started soon after Bolwig left us.

The expedition consisted of Dr. John T. Emlen, Jr., Professor of Zoology at the University of Wisconsin, and George B. Schaller, then a graduate student working on his doctorate in zoology and anthropology, now one of the foremost experts on animal behavior. The two men were accompanied by their wives, who became favorite guests of the inn. The Emlens planned to return to the States in six months; the Schallers were to stay on and carry out the program of the expedition. Both Emlen and Schaller were experienced in the field and well knew that the smaller the party, the better the chances for success.

Their first assignment was to make a six months' survey of the distribution of the mountain gorilla and to select the most favorable location for an entire year's continuous observation of their life and behavior in their native habitat.

A gorilla census would, of course, have been an impossible undertaking. A rough estimate was all the two scientists could hope to achieve. After careful calculations they came to the conclusion that there were not less than 5,000 and not more than 15,000 mountain gorillas distributed over the Eastern Congo, Burundi, Rwanda, and Western Uganda, all of them in isolated pockets. This was an unexpectedly favorable discovery. Even the lower estimate indicated that there might be a larger gorilla population than they had expected to find.

The entire party went up to stay at my saddle camp. Mrs. Emlen soon had enough of roughing it and returned to the inn. Kay Schaller, however, stayed on. Dr. Emlen followed one gorilla group for seven days, always staying a day behind the group. He made no attempt to meet them. He wanted to study their undisturbed movements. These gorillas covered about a mile a day, moving in an area a mile long and half a mile wide, crisscrossing their own tracks and sleeping every night at a different spot. Schaller climbed to the top of Muhavura on three consecutive days, finding fresh gorilla spoor but failing to encounter the mountaineering apes. Under Reuben's guidance, he learned the art of tracking gorillas to perfection.

We had hoped, of course, that the APE expedition would choose our sanctuary for its year-long program. We were disappointed when they chose instead the Kabara area between Karisimbi and Mikeno in the Congo. This was Carl Akeley's classical Gorilla Paradise. The forests there were less dense than ours, with many glades and open clearings, where the animals could be more easily observed than in our sancturary. The gorillas there were also more numerous and less disturbed by natives, for the Belgians had kept the area almost hermetically sealed, and even scientists such as the two Americans were not granted easy entrance.

The magnificent results of Schaller's work certainly have proved the choice was right. He and his wife used to return often to Kisoro to replenish their supplies. We exchanged our experiences and I was privileged to share indirectly in the exciting progress of their unique work.

Two native helpers were their only companions. George and Kay lived in a hut on the slope of Mt. Mikeno, a day's arduous climb from the nearest white neighbor, a few steps from Carl Akeley's grave. Kabara is a veritable paradise and the Schallers soon felt a part of this Garden of Eden. The white-necked ravens learned to perch trustingly on Kay's outstretched arm and to be fed from her hand. There was a green meadow with beautiful flowers in front of the hut. Buffalo grazed there like a herd of cattle. A shaggy old bull, attracted by the human presence, used to come at night and sleep just outside the hut, resting his head on the doorstep. George and Kay got accustomed to this unusual watchman and did not mind the vulgar noises he produced.

They were so far away from the outside world that not a sound of the first bloody uprising by the Bahutu against the Watutsi reached their ears, though it happened just a few miles away in Rwanda. They first read about it a few weeks later in *Time* magazine, which a boy brought up with their mail.

The Schallers fortunately came down from their hill before the upheavals that followed the Congo's independence occurred. When law and order collapsed, they were safely over the border in peaceful Uganda.

Schaller returned to Kabara during a short lull in the Congo's turmoil. He found large herds of cattle grazing there and many huts being built. With African game guards he chased away some of the intruders and captured some of the cattle. But he knew this was a futile measure. This sanctuary and its gorillas were doomed.

One day in 1959 I was delighted and surprised to receive a letter from Japan. Professor Kinji Imanishi, a prominent primatologist of Kyoto University, wrote:

We have already heard in Tokyo of you and your gorillas. We have been dreaming for many years of watching the natural life of gorillas in Africa, and we have been told that you hold the secret of the Mountain Gorilla in your hand.

Very flattering, indeed! Although we had by then much valuable material to offer, I never would have dared to make such a claim! Professor Imanishi, accompanied by his colleague, Dr. Junichiro Itani, were made welcome when they arrived. They observed the apes a few times, liked what they saw, and decided to send a small expedition on their return to Japan.

Dr. Itani was a specialist in prehuman languages and the vocal communication of animals. He could reproduce thirty-two different sounds made by the Japanese monkey. He knew their meaning and he claimed he could converse with them. When reporters came to interview this exotic pair, I introduced Dr. Itani by saying, "This Doctor is a polyglot. He understands frogs and cocks. Now he has started learning gorilla language. Every time he comes back from visiting our gorillas, he has added a few new words to his vocabulary."

The reporter laughed and used this information to add a touch of humor to the interview, which was printed under this headline:

JAPANESE SCIENTIST STUDIES
GORILLA LANGUAGE IN UGANDA

It made it sound as though this had been Dr. Itani's sole purpose in coming to Kisoro.

Dr. Masao Kawai and Mizuhara San, both experts from the Japan Monkey Center, arrived in Kisoro a few months after Dr. Itani's visit. The Japan Monkey Center is an institution that breeds and keeps the Japanese Wild Monkey (*Macaca fuscata*) for medical research. There many important studies of the social organization, behavior, and ecology of the macaques are also being made. They also organized a primate zoo, where members of this order have been collected from all parts of the world.

Dr. Kawai and Mizuhara San, who stayed with us for six months, were at all times eager observers. Once they followed the same group of gorillas for a number of days. It was a family of five, consisting of one male, two wives, one adolescent, and a tiny baby. Mother either carried her baby on her back or held it at her breast. Father now and then stopped eating to cast a wary but not unfriendly look at the spectators. But whenever the two Japanese observers dared to go closer, not only the father, as is the rule, but the whole family sallied toward them in a threatening mood. Father, still young and very strong, was quite sensitive. It would have been madness not to heed his warning.

One day, however, quite by accident, this particularly touchy family and the Japanese met head-on. Both parties were equally surprised. The gorilla father became furious and charged blindly at Peter, the tracker, who was the first in line. Gorilla and bamboo trees came crashing down on Peter. The others in the party, scared to death, threw themselves on the ground. Amazingly enough, the enraged male did not follow his advantage. He released Peter and retreated.

Mizuhara San had been second in line that day and his chest was badly scratched. He proudly insisted that the gorilla's fingernails

95

had inflicted the damage. The second tracker, who had watched the incident, was sure that the damage had been done by bamboo twigs. The chaos of the moment prevented accurate observation, so we gave Mizuhara San the benefit of the doubt.

The following day, the Japanese were unlucky enough to run into the same gorilla family again. The expedition had divided into two groups. Dr. Kawai was in the group led by Reuben, who detected the animals soon enough to call a halt just a few yards from the spot where the family had retreated for its noonday siesta. The father, stretched out comfortably and resting his heavy head on his crooked arm, seemed fast asleep. Then, from the opposite direction, with no sense of danger, Mizuhara's group arrived, again with the unfortunate Peter in the lead. Before Reuben could signal them to stop, Peter had stepped on the sleeping male, unnoticed in the shade of a bush. Father jumped up as if stung by a bee. He seemed utterly bewildered. Without giving him time to fully awaken, both groups bolted helter-skelter back to the camp.

The next scientist who proposed to come and study our apes expounded, in a long letter, the unusual way he intended to go about it. He suggested that a balloon be used to locate the animals from the air, that various cement structures be built in the forest as observation posts, and that cowbells be hung around gorillas' necks to reveal their presence in the thickets. I must have made some unflattering remarks about this investigator and his wild schemes when I wrote to Professor Dart, whose reply rebuked me.

"I did not like your remarks about Dr. So-and-so. You are there to help him and any other scientist who comes, even if you don't approve of their methods."

"Of course," I replied, "as long as I am not expected to put cowbells around the gorillas' necks."

I was less than cooperative when a certain doctor of medicine came over from the Congo. He announced he was experimenting with a new serum or vaccine he had developed against jaundice.

He had tried it successfully on tame chimpanzees. Now he was ready to test it on the larger, more humanlike gorillas, before risking it on human beings. He needed at least twelve of these anthropoid guinea pigs, he told me, otherwise the whole experiment would be useless. He asked if I was prepared to help him in serving the interests of humanity.

I protested that the gorillas were protected. "Yes, I know," said the doctor. "They are protected against hunters, but surely not against scientists. I don't want to kill them. I only want to inject them. We could easily shoot the vaccine into their stomachs with an air gun or a peashooter."

"Yes, you probably could," I answered, "but not with me! Who is going to watch their reaction? Check their pulse and take their temperature?"

That was, of course, long before tranquilizers came into use. The doctor saw my point and dropped the bizarre idea.

14

The Mountain Gorilla Observed

A lover of gorillas, I am an amateur in the true sense of the word. I did not search them out in their mountains with the passion of the explorer and the cold reasoning of the scientist, as my Japanese friend, Professor Imanishi, said he did. Yet the commercial motive and the curiosity that first drove me to the gorillas soon deepened into a genuine interest in them and in their way of life. These gorillas became my main concern. My conversations with my guests at Travellers Rest invariably focused on the fabled dwellers of our mountains, who were so much like humans and around whom there existed more myth and romance than around any other animal in Africa.

I can report only what we ourselves saw and experienced in our own particular surroundings, realizing that observations in other habitats might have resulted in different findings.

The largest unit of gorillas in our range consisted of about twenty members and was probably a temporary combination of two separate groups, since our groups were usually much smaller than that. In other habitats, however, units of up to thirty have been observed. We had one set of father, mother, and child, another of one husband, two wives, and two children. Each group kept to itself and seemed to represent a legitimate family in an otherwise polygamous society. Some groups we observed consisted of between six and twelve members, and there were always more females than males and a number of children of all ages.

Each unit had a leader, usually an old silverback male, who was a virtual dictator. He decided when to proceed and in which direction, where to stop for the noonday siesta, and where to put up for the night. When the leader went to bed the others had to follow suit. In the morning this chief gave the signal to rise. He often moved on without uttering a sound, confident that the others would follow obediently. We once watched such a silverback encouraging a lagging female to get going by giving her a solid smack on her behind.

The leader's authority is unchallenged, but the females, assisted by the younger males, are responsible for discipline in the nursery.

We were never able to determine whether there is a kind of hierarchy among the females or whether the leader has a favorite wife in his harem. We never noticed any signs of jealousy among the males. George Schaller once saw a young bachelor, who used to live alone on the periphery, approach a female in the center of the group and make love to her. The Pasha, a gorilla in his prime who could have made mincemeat of the insolent rival, just sat there scratching himself, totally uninterested in the goings-on.

Gorillas are peaceful and tolerant by nature. Ours showed no obvious signs of hostility toward each other, although the groups, as a rule, did not mix, and the pleasant habit of visiting did not exist. When two groups happened to meet, they would either ignore each other or utter a grumpy grunt, possibly as a greeting. Sometimes two groups would even stay together for awhile and then separate, each group going its own way. Schaller observed some daring members of one group getting into the middle of another.* The unwelcome intermingling caused a tremendous

* George B. Schaller, *Mountain Gorilla: Ecology and Behavior* (Chicago: University of Chicago Press, 1963).

upheaval. There was much barking, screaming, and drumming on both sides. But it was a mock fight, and not one drop of blood was shed. It seemed to have been another demonstration of the "territorial imperative."

Watching them feeding or resting, one always had the impression of a happy and friendly atmosphere. The Old Man good-humoredly endured the pranks and gambols of the lively youngsters, as they cheekily crawled and climbed all over him, and rode boldly on his back or shoulders. Only when they overdid it did he mete out rough punishment. Schaller witnessed one elderly male pulling one of his wives by one leg down a slope, in fun, it seemed, but none too gently, and then leaving her lying like an abandoned toy. Mothers with babies preferred to stay near the protection of the leader and were often seen leaning confidently against his mighty back.

When a leader gets old it seems he resigns and hands over the reins to a younger male, presumably his eldest son. The stories of competing males fighting for supremacy within their own group have never been substantiated. Frank G. Merfield,* hunter and collector in the Cameroons, tells the story of an old leader losing an arm in such a fight. Victorious, the one-armed ruler continued to reign with undiminished authority.

We had one lone male in our range who, according to Reuben, had once been a powerful chief. I doubt that the way of the gorilla is to cast out or kill a chief who has become old and weak, but Reuben always made a wide circle to evade this embittered, dethroned gorilla, who was probably quite harmless.

My guests often asked if all gorillas looked alike. I have to admit that at first I thought so. Only as one becomes acquainted with gorillas do they seem as truly individual as any neighbor. I soon learned to recognize most of our gorillas by their height, cor-

* Frank G. Merfield with Harry Miller, *Gorillas Were My Neighbors* (New York, Toronto: Longmans, Green & Co. 1956).

pulence, and other special peculiarities. One had an uncommonly big mouth, another a fat tummy. One had a scar on his face, another one leg that was shorter than the other. They could also be identified by their tempers.

A fairly safe way of identifying groups was by their numbers. How many silverbacks or blackbacks, how many wives, juveniles, and infants?

One group we recognized easily right from the start. Its leader was, of course, our old friend Saza Chief. Until his untimely death we saw him almost daily and the old blusterer had no equal in the forest. We thought him the mightiest and most powerful gorilla of the region, as imposing and forceful a personality as his namesake, the *Mutwale* Paulo. He could not, however, compete with the *Mutwale* in one respect: He had only two wives and two children, whereas the real *Mutwale* had about twenty wives, and children galore. We loved them both dearly, the one down in the village, and the other up in the mountains.

The type of nests that gorillas build depends upon the materials at hand when these nomads settle down for the night. In our forests, they usually slept on the ground, making a crude nest in a few seconds. They broke leafy branches off trees and shrubs and pulled them down, trampling them into a ring or oval with a slight rim. Usually they were too lazy, but sometimes they lined the nests for comfort, using herbs, vines, or whatever was handy.

Tree nests were not common in our sanctuary because there were not enough suitable trees in our forests. Such nests as did exist were built high in the crowns of sturdy trees and were fairly solidly constructed. They looked more like eagles' aeries than the nests of terrestrial mammals. In the bamboo zone, the gorillas demonstrated real aptitude for construction by erecting nests resembling the stilted lake-dwellings of primitive man. Pliant stems were pulled down, bent and intertwined, making a springy platform about ten feet above the ground.

The nest-building doctor referred to earlier found such a structure very comfortable. He said that the only thing he did not like about it was the faint odor. Gorillas do have the filthy habit of fouling their nests. Perhaps that is why they make a new one every night. They never sleep twice in the same bed, even if after a day's roaming they happen to return to the spot from which they had started.

The gorilla's chief enemy has always been man. After the white hunters, trappers, collectors, and Swedish princes had had their fill in the 1920s, they found it difficult to pursue their murderous trade, at least in Uganda and the Belgian Congo, where the gorillas were soon strictly protected. But who could effectively control the black hunters, whose homes were near gorilla habitats? In most Congo villages in regions where gorillas live one will meet a number of men with one withered or amputated arm or leg. They will tell you frankly that it was *ngaji,* the evil gorilla, who crippled them. Some of them have undoubtedly been injured while defending their crops against bands of gorilla raiders. The majority, however, have hunted the apes for their meat, which is considered in some areas as tasty as human flesh.

In West Africa, whence most captured gorillas hail, about three to four grown ones lose their lives for every young gorilla caught by unscrupulous trappers. Some of the captives die in transit. Thus the average toll for one live specimen reaching a zoo is alarmingly high. It has perhaps meant the death of five others. The price for a gorilla on the world market used to be about $5,000. Sir Julian Huxley showed me a letter from a French mission station in West Africa, addressed to the director of a German zoo. It offered gorillas, like colonial produce, at so much per pound, fob airport.

A gorilla does not fear other animals. His strength and appearance is too awe-inspiring. Even elephants and buffaloes respect him. In our sanctuary the other animals left the gorillas alone and they in turn certainly never picked any quarrels.

The only suspected enemy among the other animals is the leopard. He might try to snatch an unguarded young gorilla, but no reliable observer has as yet seen him do it. In the Cameroons, Merfield saw a leopard jump down from a tree and walk right through the middle of a gorilla family where the young ones were running freely about. Neither leopard nor gorillas took any notice.

Gorillas are not afraid of cows, although a herd of long-horned cattle chasing a gorilla can be a formidable spectacle. Returning from Sabinyo in a heavy rain, we once saw an elderly gorilla standing close to a herd of cows, under the protection of a tree. With arms crossed, he seemed absorbed in studying bovine behavior. When he saw us, he reluctantly left the shelter and went through the rain back into the forest.

Lions roar, leopards growl, dogs bark, horses neigh, pigs grunt, cats purr, ravens croak, owls hoot, mice squeak. And parrots screech. What about gorillas? They make all of these sounds, with some others thrown in for good measure. There are about twenty different gorilla vocalizations and each varies according to the age and sex of the gorilla and the circumstances under which the sounds are uttered.

Of course, one cannot call these vocalizations an articulate language. Despite their high intelligence, the three great apes—gorilla, chimpanzee, and orangutan—never have been taught successfully to speak. Experimenters have devoted much time, patience, and ingenuity in attempts to teach them, but the results have been extremely poor. Recent experiments at American scientific centers have proved that chimpanzees are able to learn sign language and conduct quite complex coversations.

The gorilla's vocal apparatus is not made for speech. They vocalize in the throat without the use of lips, tongue, and teeth. There seems to be no coordination between these parts and the nerves that control them. A new theory, recently advanced, suggests that the larynx and pharynx of apes and other subhuman primates are so placed that they can produce sounds but not words. This also includes Neanderthal man, whom we have always regarded as *Homo sapiens,* capable of speech. The same applies to human babies. At birth, they are not fully developed; they can cry but not form words until a postnatal development adjusts the position of their vocal instruments. Dr. Allan Gardner and his wife, Beatrice, psychologists at the University of Nevada, reported in *International Herald Tribune,* May 31, 1974, that they have taught their chimp the sign language of the deaf and dumb, and that she is able to converse quite freely by this means. She even joins words, thus forming new ones through her own efforts, revealing a capability of expressing quite complex thoughts and feelings.

The gorilla's vocalization, nevertheless, effectively suffices to communicate his emotions and his needs. He can warn his people of approaching danger, warn intruders not to approach, command bickering females to shut up, and order naughty children to behave. I doubt, however, that he reasons out loud. He cannot, for example, express such ideas as: "It was cold yesterday. I hope the sun will shine tomorrow," or "Yesterday we were in Rwanda. Let us climb Sabinyo tomorrow."

Gorillas in general are a silent people, not noisy and extroverted like their relatives, the chimps. The sounds one hears most often are a contented grunting and belching while they are feeding, and an angry bark when they are disturbed, the kind of bark that seals rather than dogs produce.

Reuben and Jill once heard voices in the forest. Reuben thought he might catch some illegal woodcutters and stealthily approached the spot. Instead, he found a group of gorillas sitting under a tree engaged in a lively conversation. It had sounded so human that even Reuben, who knows their lingo and can imitate their sounds to perfection, had been deceived.

Unique and quaint is the hooting sound, which is usually the prelude to the climax of their emotional expression: the drumming display. Like his cousin the chimpanzee, the gorilla is a passionate drummer. He beats his flat or slightly cupped hands in rapid alternation upon his chest, abdomen, hips; on tree trunks; on the ground; on whatever is handy. The effect is remarkable. Gorilla drumming sounds the alarm in case of danger, and also threatens the invader. "One more step," it says, "and there will be trouble!" That drumming used to frighten me at first. Later I found the exaggerated excitement rather comical. When they are by themselves, gorillas often drum just for the fun of it. We once watched a tiny tot belaboring his miniature thorax with great effort but little result. After each roll he stopped and looked around, as if making sure that his performance was duly appreciated.

Our gorillas made a specialty of drumming by beating their cheeks. It sounded as if the drummer in a jazz band had struck his instrument's rim with his drumsticks. Reuben had actually seen this kind of drumming and could imitate it, but was unable to convince George Schaller, whose gorillas over the border apparently did not know this nuance.

One night we were drinking coffee with some guests in the saddle camp kitchen hut. We knew that a gorilla family had set up its camp on the slope behind us so we spoke in whispers. As usual, the conversation was all about the gorillas. I encouraged Reuben to show off a little and entertain our guests with his star act, "Old Silverback," a clever imitation. He started shyly and quietly but soon forgot himself, hooting and clicking full blast. Instantly from up the hill came the response, "hu, hu, hu, hu," in high falsetto, followed by the wooden "click, click, click, click." A cordial to and fro went on for awhile between original and imitator. Then an infant, apparently disturbed in its sleep, began to whimper like a puppy that had not yet learned properly how to bark. Eventually the mother's voice, trying to calm the child, joined in. A fine example of the wide range of gorilla vocalization—and an unforgettable experience for us and our guests!

15

Are Gorillas Dangerous?

The one question all our guests asked was: "Are gorillas dangerous?"

I wouldn't call them perfectly gentle; as a matter of fact, I was shaken to the core every time I came close to a gorilla. "This cannot be true!" I always said to myself. "Am I really standing here, just a few yards from this monstrous shape—with nothing between his wrath and me?"

Reuben and I, who could claim to know gorillas better than all the hunters, explorers, writers, and film producers at that time, were never tempted to appear heroic by exaggerating the dangers of our daily contacts with these animals. A full-grown male is certainly a frightening sight, particularly when he stands upright before you for the first time. His canines are terrifying. I never could figure out why nature had equipped this vegetarian with such vicious teeth! For defense, according to scientific explanation. With his large neckless head, massive shoulders, and muscular arms, his body bent slightly forward, he resembles nothing so much as a heavyweight wrestler ready to come to grips with his opponent. The gorilla usually walks on all fours, the bulk of his body borne by surprisingly short and bandy legs. The knuckles of his long arms touch the ground as he moves forward. He is not built for speed. In the dense forest and the bamboo thicket he can move faster than a man. On open, flat ground he cannot compete with us.

I think it always depends on one's approach to wild animals, whether they regard you as friend or foe. Our gorillas were not dangerous because we took care not to harass them. They probably considered us a nuisance but certainly did not fear us. All genuine observers have taken to the "gentle" apes and speak of them with great affection. The gorillas who attacked hunters and collectors were dangerous because the men were dangerous. Hadn't they come to kill or capture them?

Dangerous or not, gorillas are unpredictable. One must be prepared for surprises. To safeguard myself, I asked visitors to sign an indemnity form before going into the sanctuary, agreeing that I should not be held "liable to them or their heirs and successors for loss of life or any damage or injury they or their property might suffer." Frankly, the precaution was a little overdone, but most visitors were not intimidated. On the contrary, they seemed to feel that danger enhanced the thrill. As they signed the form, which they received later as a souvenir, they were obviously enjoying in advance the admiration such undeniable documentation of their bravery would arouse back home.

Sometimes visitors wanted to turn back en route when they found that none of us were armed. The panga, which the trackers always carried for cutting the path, could in an emergency serve as a defensive weapon. I, too, usually had a panga in my hand and I had learned to use it effectively. I am very sure that I would not have let myself or any member of my parties be torn to pieces by a gorilla.

But clearly my primary concern was to take care of the apes. As Honorary Game Ranger, I was officially responsible. I still shudder at the thought of what the Lord Protector in Entebbe would have done to me if I had been forced to use my weapon in an emergency. The slightest harm done to a gorilla would have had serious consequences. On the other hand, nobody would have been concerned very much if a gorilla had harmed us.

I am thoroughly convinced that wild animals will not attack unless they are provoked or have had previous bad experiences with humans. When an animal is attacked, or when he thinks he is in danger of attack, he will bravely defend himself and his family. But I know of no other wild animal in Africa of the size and strength of the gorilla whom one could approach so closely without great risk. I always suspected that my gorillas possessed a grumpy sort of good humor and benevolence, kept hidden behind their forbidding appearance and fierce ado.

I know of only one case of a gorilla actually killing a man. The animal was shot afterward and the medical report described it as a barren female. The natives had known of her living alone in the forest. Restlessness, it seemed, had driven her down into the plain. I am convinced that she was half mad with loneliness and frustration, and that this is why she attacked the first living thing she encountered.

Carl Akeley ascribed the gorilla as "an amiable and decent character." But he also cautioned that "the white man who will allow a gorilla to get within ten feet of him without shooting is a darn fool!" I think he was wrong there. Reuben, with me directly behind him, repeatedly stood much closer to gorillas who were by no means amiable. We did not shoot and we are still alive.

No hunter, explorer, or collector I know of ever willingly spent the night close to a gorilla. George Schaller, however, once decided to spend the night near a gorilla camp in order to observe the family when they got up in the morning. Schaller slept peacefully, but when he awoke he heard heavy breathing coming from the other side of the bush under which he lay. A few yards away, a little too close for his liking, a hefty male gorilla was still fast asleep. George had not noticed him in the darkness when he had cuddled into his sleeping bag. He took care not to disturb his roommate and silently stole away, realizing one can never tell what a gorilla will do when disturbed in his sleep.

Schaller made many of his studies in open woodland. The

gorillas almost always could see him coming and could watch him climb into a tree in full view. Females repeatedly climbed the tree from where he was watching the family. Once a daring mother with a baby at her breast even shared the same branch with him.

In our dense forest, however, they could not see our approach. They only *heard* something approaching, something that could be an enemy. They were understandably nervous, always prepared to defend themselves and to frighten any intruder away. I took care not to frighten or surprise them and we were able to come very near.

When we met them in the open with fifty or a hundred yards or a deep ravine between us, they would continue whatever they were doing, feeding or resting, with complete unconcern. But when they heard us coming, without being able to see us, old silverback would receive us with one of his acts of ferocity.

The golden rule on such occasions was never to turn and run. If retreat seemed advisable, it had to be in slow motion. Hasty movements are always taken as threats by animals, even domesticated ones. If we stood firm, our opponent would simply stare at us and we would stare back at him.

The gorillas even practiced that staring trick among themselves. It seemed to demonstrate submission and friendly feelings. The object of the stare was expected to turn his head sideways and shake it rapidly back and forth. We never tried to do this. We simply stood our ground, knowing that the gorillas were at least as afraid of us as we were of them. We knew that the intimidating stare of the old one was pretense and that in the end it would be he who would waver and retreat.

But as long as there are rules, people will make exceptions to them. One day the Chief Game Warden came to photograph gorillas. Reuben found a fine, full-grown male to pose for the bwana. However, the light wasn't perfect, and while they waited the model noticed them and moved off. The warden and Reuben, disobeying all our rules, crawled after him. Reuben, at least, should have known better than to follow a gorilla after an encounter. Such pursuit was bound to give him the feeling of being hunted.

The warden himself should have respected his own instructions. But, no! The two crawled after the gorilla through a tunnel of wet vegetation. Reuben, in the lead as usual, went right into the ape's open arms, but was not received as a cousin. The gorilla closed his hands tightly around Reuben's throat and began to strangle him. Reuben thought he was a goner. Suddenly a shot rang out! The warden's gunbearer had promptly fired a shot over the gorilla's head. The strangler was shocked. He stopped short, let Reuben go, and scrambled panic-stricken up the hill.

A shot in the sanctuary, where all firearms are banned! Apparently the ban did not apply to the Lord Protector of Uganda's wild animals. We forgave the gunbearer his weapon, for it had saved Reuben's life and probably that of the Game Warden as well.

Imagine the Game Warden's surprise, then, to see Reuben in hot pursuit of his attacker! He shouted after him, "Reuben, are you mad? Come back!"

"That bastard took my kepi," Reuben shouted back.

"To hell with your kepi! I'll give you a dozen new ones, but for heaven's sake, come back!" the Game Warden yelled. Reuben regretfully obeyed.

At that time, Reuben was not yet employed by the Game Department and therefore was not entitled to a Game Ranger's uniform. To bolster his prestige as a guide I managed to get some odds and ends out of the department, including a sweater with GAME embroidered on the chest and, best of all, the treasured kepi. This wonderful headpiece had a neckflap and it gave Reuben the appearance of a Foreign Legionnaire. It also sported the Game Department's badge, of which Reuben was justly proud.

I must confess that I sometimes became a little vague about my answer when guests continued to ask, "Are gorillas dangerous?" When I broke my leg in 1957 and was hobbling about on crutches, every visitor quite naturally assumed that I had been in a melée with a gorilla, and that only my strength, presence of mind, and skill in judo had saved my life. I was considered a hero. Should I be blamed that I did nothing to destroy the legend?

My friends knew, of course, that I had broken my leg jumping over a lily bed in the garden.

16

How Human Are Our Cousins?

How close is our kinship to gorillas? How human are our cousins?

I myself have never regarded them as mere animals. I always felt a close affinity to them. "They are really wonderful, these gorillas, so friendly and interested," George Schaller wrote to me shortly after he had started meeting them.

Alan Moorhead, the renowned author of *The White Nile* and *The Blue Nile*, arrived unexpectedly at Travellers Rest one night. Visiting gorillas had not been on his program; he was prompted only by a last-minute curiosity. Later he was to write in my guest book, "To see the gorilla, as I did today, must surely be one of the rarest and most exciting experiences in Africa." Moorhead related a moving account of his meeting with "A Most Forgiving Ape," as he called the chapter in his book, *No Room in the Ark*.

"Oh, my God, how wonderful!" he remembered calling aloud at his first glimpse of Saza Chief. "He was glaring fixedly upon us and he had the dignity and majesty of a prophet. He was the most distinguished animal I ever saw and I had only one desire at that moment, to go forward toward him, to communicate."

Gorillas are like us in so many ways. They live and die, copulate and reproduce like us. They get sick from the same diseases as those we suffer from. They belch, cough, hiccup, sneeze, yawn, pick their noses, and break wind, just as humans do. They love, care for, protect, and discipline their children. They like and love one another. Mother love, in particular, is very pronounced, as it is with all mammals. Love is an essential emotion in the life of these apes. You cannot help feeling it as you watch them. They cannot bear loneliness, as experience in zoos has proved. If there are no visitors in the zoo, the animals mope. In order to entertain them, TV sets have been installed in the Dublin zoo. The apes are thrilled by Westerns and documentaries featuring animals, but bored with news and politics. They are even less enthusiastic about police stories, and one gorilla becomes enraged every time he sees Kojak on the screen.

A gorilla in the wild leads an ideal life, a happy-go-lucky alternation of eating and sleeping. He has nothing to worry about. He goes to bed with the sun and rises with it. He roams his domain at leisure, feeds while he walks, has a good rest at noontime, and never overexerts himself. Unlike the chimpanzee and other primates he dwells mainly on the ground. He can climb trees and build nests in the branches, but he lacks the agility of a good climber.

Now and then in our tracking we would come across a single nest, a silverback's, to judge from its size. It was always a few hundred yards away from the family but within call in case of danger. Reuben was inclined to interpret animal behavior in human terms and he suggested that the separate nests indicated that the male was shunning the female's presence during her pregnancy.

Gorillas are slow breeders, much slower than man. No pill is required to control their birth rate. They do not produce one child a year as do many human families. They wait three to four years until one infant has achieved independence before they give birth to another. That is probably why most gorillas have more than one wife. Most primitive societies also practice polygamy. A native African woman used to give birth to a baby about every three years. By inculcating monogamy, the Christian missions changed all that.

The result is a skyrocketing birth rate—one baby per woman per year.

Although special breeding seasons have been observed in other primates, no particular period exists in the life of the gorilla. They are born in any month and evidently gorilla females are sexually receptive throughout the year.

In the wild, sex seems to be less important to the gorilla than sleep and food. In captivity, where there is a lack of exercise and no problem of finding food, sex is practiced more actively and starts at an earlier age.

The mating acts that Rosalie and I observed were performed the human way. With captive gorillas other positions have also been observed. Rosalie's incident took place in full view of the whole group with a youngster swinging on a creeper to and fro above the mating couple. The scene I witnessed occurred in the privacy of some greenery, a little bit away from the family.

Until recently gorillas never reproduced in captivity. During the last fifteen to twenty years, however, some zoos, with much care and understanding, have succeeded in breeding them. The first was in Columbus, Ohio. The parents were Western Gorillas: Baron, the male, was ten years old; Christina, his wife, six. Christina had tried to attract Baron's attention since the tender age of five. The "young man next door" had not been particularly interested at first, but gradually he became an ardent lover, so stormy, in fact, that the lovebirds had to be separated. Seven months later Christina produced a premature baby, which had to be kept in an incubator.

Since then Western Gorillas have produced progeny in the Basel and Washington Zoos, among others. The first Eastern Gorilla to be born in captivity was born in the Antwerp Zoo. While twins have never been observed in the wild, the first set in captivity was born at Dr. Bernard Grzimek's Frankfurt Zoo.

Mothers of gorillas born in captivity do not seem to know what to do with the little stranger that has emerged from their wombs. They remain indifferent, show no love or care, and do not protest when the baby is removed from their cages. It seems that maternal instinct is not hereditary in anthropoid apes but must be acquired by experience. The most famous case perhaps is that of Pattycake in New York's Central Park Zoo. In 1973 both parents fought over the newborn baby and broke her arm, which had to be set in a hospital. It was several months before Pattycake could be reunited in the cage with her parents.

Goma, the first gorilla born in the Basel Zoo, had to be taken away from its mother and reared in the house of the curator, Dr. E. M. Lang. Without having to wait three years for her baby's independence, as do mothers in the wild, this mother had a second child, Jambo, a year later. She took this one tenderly into her arms, but had to be shown how to hold and nurse it or it would have died of starvation. After that she became a good mother.

Gorillas also resemble us anatomically, but have a longer appendix and a shorter penis. *Homo sapiens,* has of course, a bigger brain and the largest penis of all 193 primates. The sexual organ of the gorilla, the biggest and mightiest of all primates, is surprisingly small. That makes it a tricky business to determine the sex of the young ones.

Belle Benchley, the former director of the San Diego Zoo, describes in her charming book, *My Friends, the Apes,* the daily lives of "Johnsons' black babies," two gorillas which the zoo purchased from Martin and Osa Johnson, the famed early explorers of African wildlife. The "black babies" hailed from the eastern Congo and were the first mountain gorillas ever to reach a zoo. They were about four years old when they arrived and were thought to be a pair. Their behavior puzzled experts at the zoo for several years until they discovered that both were males!

The Basel Zoo was proud of its young male gorilla named Achilles. Only when he had an operation after swallowing a pen given to him by a stupid visitor was it revealed that Achilles would have to be renamed "Achilla."

In the wild, food seems to be more important to the gorilla than sex. The greater part of the gorilla's day is spent eating. Their table manners are impeccable. Gorillas do not graze or browse, or stuff their stomachs to capacity, then regurgitate, as most vegetarians do, but sit down comfortably for their meals. They either help themselves with their long arms to whatever vegetable is near at hand, or they first collect their menu, squat down at a convenient place, and eat what they have gathered. When they move on, they leave behind the remnants of their meals in neat piles.

Reuben once saw a gorilla hold his cupped hands up to some water that was dropping from vegetation above him. Another time he watched a gorilla drinking from the watercourse. He did not bend down as a man or other animals would, but ladled it out with the hollow of his hand.

I have known other observers who have seen gorillas make knots in creepers to facilitate climbing trees and use sticks to reach honey in a beehive, inaccessible any other way.

It is unfortunate that no camera was handy to record these incidents. Photographic evidence would have delighted the American tool manufacturer, Leighton A. Wilkie, a visitor to Travellers Rest. We knew that he was keenly interested in the history of the development of the tool, especially in information indicating that certain animals made and utilized rudimentary versions. A contributor to Professor Dart's research, before Dart's fellow scientists were ready to accept his findings, Wilkie was eager to learn whether our gorillas had been observed using sticks and stones as tools. It is only a short step from drinking from the hollow of the hand to using a cup, from licking a stalk to using a spoon, from knotting creepers to making a rope ladder.

Wilkie was disappointed, and so were we, that we could not produce the evidence he wanted. In fact, experiments with captive apes indicate that the gorilla is the least technically minded of the three great apes. The teachable chimpanzee is mainly a clever imitator, but he can learn to perform amazing feats involving the use of tools. The brooding orangutan, the only animal known to make tools as the need arises, has amazing dexterity and ingenuity and is undoubtedly the most mechanically gifted of the apes.

There are exceptions to the rule, if certain reports of the achievements of individual gorillas are to be believed. Friends in Ghana once wrote me about a female gorilla, kept in the garden of a hotel in Accra. When they brought food to this ape, they could not get near enough to the bars to feed her by hand; they had to offer fruit on a stick. She let the food drop carelessly, but she kept the stick and used it later to pick up tidbits thrown into the cage beyond her reach.

It would seem that the gorilla is not lacking in intelligence. He is, however, heavier and slower than the nimble, extrovert chimp; he is an introvert of great dignity. And, if he is not as inquisitive as the chimp, he seems to possess a more solid kind of curiosity.

An Italian, building a road through the forests of Mount Kahazi in the Congo, once told me about a gorilla family that, despite the noise being made by men and machines, emerged from the forest and, for more than half an hour, quietly stood watching the activities on the busy road. They seemed to be fascinated by the performance of the tractor.

17

Death of a Chief

Sometimes guests at Travellers Rest were provided with extraordinary experiences. One morning Reuben took a party up to see the gorillas. Saza Chief and his family had been living in the lower forest at that time and I thought I could depend on the old ruffian to perform in all the ways that could be expected of a silverback of his stature. I took this so much for granted that I had stopped speeding out to ask returning parties whether or not they had seen the gorillas. That morning, however, the party returned so early that I was curious and hurried out.

"Back already?" I asked. "Did you see them?"

"Yes, we saw one, a dead one," was the answer.

A dead gorilla? That hit me harder than I can express. Instinctively I felt that the dead one must be my friend, Saza Chief. We knew he had been wounded a short time before in a fight with another silverback, one older and even stronger than he. But dead? That would be a sad loss indeed.

It would also be a momentous event for a student of gorilla behavior; it might even provide an answer to the question: "What do gorillas do with their dead?" I had never believed in the fabled cemetery of the elephants, and the idea of a gorilla graveyard had seemed equally fantastic to me. However, I had often wondered why we had never come across a dead one. Considering the vast terrain over which the few groups of gorillas were spead and their nomadic way of life, perhaps this was understandable. Most free-living animals, feeling death near, look for a hiding place where they can die alone. The dampness of these forests accelerated the process of decomposition, and hyenas, jackals, and other carrion-feeders saw to it that, in no time, nothing of a corpse remained.

Reuben apparently had found such a hiding place. He had found the body of a dead gorilla, one that had died a natural death. Schaller, Fossey, and other observers later found quite a number of such corpses but Reuben was the first to make such a discovery and he had made it only by chance.

Twelve days earlier Reuben and his trackers had heard the din of battle in the forest—a terrific barking, roaring, and screaming. The men had approached the place of the uproar and, through a curtain of greenery, had gained a furtive view of the scene. The two antagonists were biting, scratching, and beating each other so violently that they were unaware of the spectators. Reuben thought it wiser to retreat for he feared that if the human intruders were perceived, the fighters might forget their own quarrel and make common cause against the human enemy.

Reuben took me the next morning to the battleground. A spoor of fresh blood led high up into dense scrub. Tracks of blood are not common in nature, at least not in protected areas. They always call for investigation. In this case the problem proved simple: the blood was gorilla blood. It marked the trail of the two combatants, and it indicated that one of them had been badly wounded. We knew he could not be far off but Reuben was taken wholly by surprise when, a little higher up, a silverback suddenly broke from a bush and charged him.

Accustomed to that sort of bluff, Reuben was at first not perturbed. We remained standing and glared at the ape. This time, however, the Golden Rule did not apply. We should have realized that a lone, wounded gorilla does not behave in a normal way, and that the bluff of a nervous, pain-racked animal might become deadly earnest. The situation called for quick judgment and Reuben's sure instinct prompted him to do exactly the right thing. He stood his ground and waved his arms about. He screamed and barked back at the wounded male with all his might. But his voice sounded like that of a hoarse cock in comparison with the gorilla's roar and if the situation had been less precarious, Reuben's performance would have been extremely funny. Yet his boldness did the trick: The enraged one stopped short of a second charge and returned angrily to the bush.

"Saza Chief," Reuben had whispered uneasily. I, too, thought I had recognized our friend in this bleeding, disheveled figure. Our mighty chief had been critically wounded.

Nothing moved in the forest. There was not a sound. Both Reuben and I sensed that a fierce battle had just ended close by. The trampled vegetation betrayed the recent presence of other animals watching the fight. As we followed a track downhill, we came upon two females and two youngsters. Without a protective male hovering nearby, they were clearly the family of the wounded chief.

Some days later we found, at another spot of flattened greenery, tufts of gorilla hair in large quantities. Had the two opponents clashed here again? Once again, we met the dejected family, still unprotected, wandering aimlessly through the forest.

We also found the camping site of the victor. Judging from the number of nests, this new group seemed to consist of one male, five females, and one child. They were apparently fresh arrivals, who had just crossed over from Rwanda. To judge from the little that Reuben had seen of the fight, this new male was a powerful figure and a tough adversary even for our mighty Saza Chief.

The quarrel obviously had not been a simple test of strength between two males competing for supremacy within a group. Neither was it a struggle for possession of the females, for the victor had made no attempt to take over the family of the loser. In my opinion, we had witnessed a territorial challenge, Ardrey's *Territorial Imperative,* a competition rare among gorillas, though common among most other animals. Saza Chief had long been undisputed master, Lord of the Lower Forest. No other group had dared to enter his territory, but now these immigrants from Rwanda had invaded his Lebensraum. There was enough food and space there for both groups, but gorillas are not sociable. Saza Chief must have felt it necessary to defend his kingdom—and he had failed.

Eleven days after the first battle, the forest reverberated again with sounds of war. The screams and roars were terrifying, of such untamed elemental ferocity that this time even Reuben did not dare to peep through the curtain of leaves behind which the battle was being fought.

The next morning, when Reuben took his party to the battlefield, the forest was as silent as death, the air thick with disaster. The scene was a chaos of torn-off leaves and vines, of broken branches and flattened vegetation, but not a drop of blood could be seen. As if in a state of hypnosis, Reuben walked to one of the bushes, bent its twigs aside—and stopped short. He beckoned to the others to come. They told me later they all were moved by the majesty of death.

We had known for days that our friend would be the loser, but we had not reckoned on his death. He will escape to Rwanda, we reassured ourselves. He will lick his wounds there and return some day.

On receiving Reuben's report, I immediately sent him up again with sixteen porters to fetch the body. Because my broken arm was still in a cast, I thought I would await their return at the foot of the mountain. But people gathered there, so I followed the porters. I did not want to pay my last tribute to Saza Chief, my good friend, in such a crowd.

When I reached the spot, I was glad I had struggled up to see him. There he was, the mighty chief, stretched out on his back, legs drawn up and hands twisted in agony. One eye was swollen shut. No one had closed the other one. No one had been near him in his last hour. I was deeply moved.

The corpse was tied with creepers onto two poles. Four men on each side, sweating under the heavy burden, staggered down the steep slope, roughly letting their load drop when they arrived at the Gombolola Headquarters. The chief there permitted me to use the Village Hall as mortuary. I asked the porters to carry the body to the steps and set it up in a dignified position.

Saza Chief had meant much to me. I felt I owed him some respect and directed that his silver back be set against a wooden pillar. His mighty head was sunk on his breast, his mouth, with the terrific canines, was open as if he were crying out in pain. His enormous, yet so touchingly human, hands were closed to fists. The wounded eye was shut and the open one looked with sad disdain at the surrounding mob.

The arrival of a dead gorilla could not be kept secret. The whole of Kisoro and its environs, from babes on their mothers' backs to old Methuselah, streamed into the village. Even from across the border they came—Europeans, Africans, Indians. Who could blame them for their curiosity? There was no cinema anywhere, no horror film to make their blood curdle, so our poor old Saza Chief had to play the role of Frankenstein's dead monster.

I quite understood why people wanted to have a close look at the rare sight of a dead gorilla, but it was beyond my comprehension that European and Indian mothers—the Africans were more reserved, almost indifferent—brought their tiny tots to view that corpse. Despite its touching humanness it was a ghastly sight.

After the mob had dispersed we carried the corpse into the hall. Leaning against the bench, looking defiantly down upon me, the dead chief appeared alarmingly human in the dim light. Man or ape? What keeps us apart? Lost in the sight of this mystery, I felt as though I were looking back over millions of years, back to the very beginning of time.

It took much persuasion to convince the African sergeant in

charge of the police transmitter that a dead gorilla was important enough to justify sending a message to the Game Department in Entebbe. I knew that the corpse of a gorilla would be of considerable value to the Medical School of Makerere College. I asked the Game Warden to arrange for some experts to come immediately and tackle the delicate job of skinning and dissecting. But I had to be prepared for an emergency. To Reuben and his men the dead gorilla was taboo. They refused to touch it. Such unclean jobs were left to the Batwa; it was they who had carried the corpse down.

I knew that it was necessary only to remove the important organs undamaged and preserve them. The ornithologist of the Coryndon Museum in Nairobi once had shown me how that is done with birds and bats, but I did not know how to go about the more formidable task of a gorilla. The corpse was decaying rapidly. If the experts did not come soon the unpleasant business would be up to me. That thought haunted me all through the night.

I waited until noon. Something had to be done. The body was now stinking. We were sharpening our knives when a van from Kabale arrived to collect the corpse. A professor of pathology happened to be staying at the White Horse Inn there and, though it was outside his own sphere, had agreed to take over the dissection. With a sigh of relief, I watched the hearse speed away.

When the body was dissected that evening it was found that Saza Chief, despite his great loss of blood, had died not from his wounds but of asphyxiation caused by regurgitating food which had gotten into the windpipe. Neck, throat, and gullet showed severe bruises, obviously caused by strangling. The pathologist certainly did his best, but he was criticized for not having saved some valuable parts.

The police even frowned at me for not having taken the dead gorilla's fingerprints! I wish I had. They might have proved quite instructive. I remembered that the hands had shown crisscrossing lines. The lines of life, heart, and destiny were all there, just as in human hands. An expert palmist might have read from them the age and character of the deceased.

In 1946 a fine specimen of a male mountain gorilla had been collected in our range for the Coryndon Museum. There, mounted by an expert taxidermist, it makes a good display. Our own collection of Ugandan mammals in the Kampala Museum could not pride itself on possessing a gorilla. I had hoped that Saza Chief, mounted in a characteristic posture, would fill that gap. However, the taxidermist would not accept the job since inexpert handling had damaged the skin. The pelt had lost too much hair, and fingernails and eyelashes were missing altogether.

They returned the skin to me—and what on earth was I to do with it? I thought of having it prepared, but I was out of funds. The idea of having Saza Chief hanging on the wall of my lounge or lying as a rug by my bedside did not appeal to me. I carelessly stored the evil-smelling pelt in an open shed and was to have reason to regret my carelessness. Susi, our pig, to whom nothing edible was sacred, discovered the tidbit and feasted on it. When I surprised her, nothing but scraps were left of what once had been the lord of the forest.

These remains were interred with all due respect in our cemetery of pets.

18

Reuben Gains a Namesake

About a year after the death of Saza Chief, Reuben discovered another gorilla corpse in the same forest. He had expected this new discovery because he had realized that the victor who had downed the chief was now himself in bad shape. The new lord of the forest had not enjoyed the fruits of his victory very long before he had been smitten with a terrible disease which, starting like a harmless cold, had turned into acute diarrhea, making hell of the old ape's life.

This spendid silverback had never had a masterful personality. He was not the born gorilla ruler that his predecessor had been. He was reserved and undemonstrative. Mock attacks were not his line. When we encountered him, he would give a halfhearted bark, a hesitant grunt, then quietly disappear.

Suddenly, his whole disposition had changed. Reuben, accompanying the Game Warden, encountered the sick male with his child. He may have felt cornered, for he became threatening. Reuben took the hint. This time he and his party quietly withdrew.

The silverback and the child became inseparable. Such a close father-child relationship is not uncommon among gorillas. What was remarkable was the fact that the child's mother along with the four other females always fed apart. Perhaps this was because the sufferer fouled the air with a pestilential odor. Were the gorillas practicing a crude sort of hygiene so as not to get infected themselves? But who had determined that the child would stay with the father?

There was nothing we could do for the sick ape, but Reuben visited him daily. His condition grew worse and we were prepared for the end.

Reuben's party found the child crouching on the dead father's shoulder. There was no sign of the females, although the day before they had been feeding nearby. They had abandoned the dying husband and also the helpless child. Left alone in the forest, the little one would probably have died of sheer loneliness or fallen prey to hyenas, jackals, or leopards.

Reuben cautiously approached the corpse. Every time he tried to grab the youngster, it would bare its teeth. The others in his party, even Peter, the tracker, kept a safe distance, but Reuben kept trying. Eventually the little fury jumped off the dead ape's shoulder and scampered off full speed into the brush with Reuben right after him. After a hard struggle—and a bite in his shoulder—Reuben caught the amazingly strong youngster in his groundsheet. Exhausted from the fight, the little gorilla offered no further resistance and soon fell asleep while being carried down on Reuben's back. He awoke only when he felt the motion of his first drive in a jeep. Then, cleverly freeing himself from the rope tied around his neck, he again attempted to escape from the groundsheet.

I had not been prepared to welcome the wildly kicking object that emerged from Reuben's groundsheet.

"Boy or girl?" I asked Reuben, knowing full well how difficult it is to tell with young gorillas. Reuben did not need three or four years to find out, like the experts in the San Diego and Basel Zoos. He cast one short glance at the little ape and announced, "It's a boy!"

He was about two years old, this handsome gorilla lad. He was the size of a medium poodle but much heavier, weighing about forty pounds. In honor of his savior I called him Reuben, Jr. At first Reuben, Sr., was very pleased with the name but later he asked me to change it. The villagers were making fun of him for having a gorilla as namesake.

Where were we to put up our unexpected guest? We tied him to a tree at first, but the crowd of spectators made him nervous. He was unfriendly; no one could touch him. So we quickly prepared an old rabbit hutch as a temporary lodging, and Reuben managed to transfer the violently squirming bundle into his new quarters via the groundsheet again. By that time he was in such a state of hysteria that we thought it best to leave him in peace. I covered the hutch with the groundsheet, so that he could recover from the shock in privacy. Two guards, armed with pangas, had to be posted by the hutch to keep away people who thought we were making too much fuss about that little monkey.

After awhile I carefully lifted the cover and talked to the frightened creature, huddled in the farthest corner with his back to me. My first "hu! hu! hu! hu!" in gorilla falsetto had no effect. Then he could not resist opening an eye and casting a furtive glance at me, quickly hiding his sad, handsome face again in his hands.

Feeding was less difficult than I had expected. Like all children, our young ape ate the whole day long. Although apes refuse all food when they are emotionally upset, our infant soon exhibited a healthy appetite. As long as I did not watch him, he would relish the wild celery I had brought down from the forest especially for him. As soon as I lifted the cover, the eating would stop. Once I surprised the little fellow while he was biting into a stalk of celery. As if to annoy me, he pushed the whole thing angrily into his mouth like a naughty child, shooting furious glances at me as he did so. Later he ate bananas, carrots, sugar cane, even bread and *posho,* and drank from the bowl of milk I managed to get into his hutch without being bitten.

To retrieve the body of the little fellow's father, I had to collect a squad of porters again. It was already dusk when we arrived at the spot where the long-suffering gorilla had given up the ghost. Death had come to him as a friend; the relaxed body with open eyes rested on a bed of greenery and showed no trace of the recent suffering. The mouth, despite its ugly, decayed teeth, seemed to be smiling.

This time I took care not to have the body tied so close to the poles that it would rub against them and be damaged. It was dark when we arrived at Gombolola Hall. No curious mob was waiting this time.

The radio sergeant was gorilla-conscious by now, and I had no trouble getting a message transmitted to the medical school at Makerere. I expected a team of experts to arrive the next morning but they failed to appear. The body, in perfect condition, had to be skinned and disemboweled without them. Once again, much of the body was lost. Skull, skeleton, and the most important inner organs were saved. The vet in Kabale undertook the post mortem and stated that death had been caused by "gastroenteritis."

Reuben, Jr., soon would have settled down at Travellers Rest and adopted me as his second father, but I felt I could not accept the responsibility. It would have been a full-time job to give him the love and care and constant companionship that young gorillas need. They are inclined to be melancholy and fade away when left to themselves. Moreover, such creatures are delicate in captivity; without a hospital or doctor nearby, there was no one who could have taken care of Reuben, Jr., if he had become ill.

I knew, too, that sooner or later I would have had to part with him. At a public place like mine, it would have been risky to let him run about freely. Most guests—especially the children—would not have known how to treat a pet of such strength. Tempestuous in their affection, gorillas can easily break ribs and collarbones in a loving embrace. The guests might have rubbed him the wrong way and I would have been held responsible for the damage. Nor would I, who was accustomed to seeing gorillas living free in their forests, ever keep him in a cage.

Reuben, Sr., was disappointed and angry with me when I allowed his namesake to be taken by Landrover to the Game Department in Entebbe. But I knew it was the best solution for the gorilla and for me. The Game Department kept an animal orphanage there and Cleo, a one-year-old chimpanzee, an ideal companion, was waiting for our little orphan. The two were to share the same cage.

It was love at first sight. Reuben was always jealous when the keeper entered the cage and tried to flirt with his little girl friend.

Thanks to Cleo's influence, the drastic change of diet, usually a critical period with captive gorillas, was no problem at all. She made her friend eat and he thrived. Strangely enough, he rejected all the forty-odd varieties of bamboo shoots—the only familiar items on the menu—which were native to the region of Lake Victoria, where Entebbe is located. Instead he acquired a taste for exotic fruits, such as papaws. They stimulated his bowels to such an extent that it was feared he might develop the same sort of enteric troubles from which his father had died. Fortunately, the pith of elephant grass, fibrous and rich in salts and proteins, soon cured him. While old gorillas are fussy eaters, preferring to starve rather than try anything new, youngsters respond to any reasonable diet. Today they are even fed meat in some zoos.

Young Reuben soon lost his shyness and became the most popular member of the orphanage. Many visitors came to admire the two charming little apes. I would have liked him to remain in Uganda, but the Game Department did not feel up to shouldering the responsibility for such a valuable animal. Like me, they also needed money, so it was decided to sell Reuben and Cleo. I suggested that he be sent to Regents Park Zoo in London. Their collection lacked a mountain gorilla, and I felt that our Reuben, as a Uganda citizen and member of the Commonwealth, should go nowhere else but to Britain.

He was sold for fifty thousand shillings, a handsome price at the time. Fifteen thousand was given to the village of Nyarusiza. It was hoped that the money would go toward educating the people on the value of these apes, convincing them that it would pay to preserve them. I was a bit doubtful and thought that the policy might have the opposite effect—to encourage the natives to catch more apes and do a roaring trade. Reuben and I received generous gifts for our trouble. Reuben used his windfall to build himself a better house.

On young Reuben's arrival in London the rascal made himself unpopular with the press by bombarding reporters with straw. Reuben and Cleo were very happy at Regents Park Zoo. They were parted when the zoo acquired two young Eastern Lowland females. The zoo hoped to breed the gorillas in due course.

It saddens me to report that Reuben, Jr., died of bronchitis and pneumonia during the severe winter of 1962. His two mates died also. I learned too late that the primate house in Regents Park apparently could not be heated sufficiently at that time. Gorillas used to catch pneumonia and tuberculosis and die prematurely there. They now have a fine new ape house: it is hoped that gorillas will survive and breed there as they do in other zoos. I could never understand why our Reuben, who was accustomed to a wet, cold climate, could not withstand the English weather. The magnificent Guy, the biggest and oldest captive gorilla in Europe, hails from the hot west coast and has been living and thriving there for over twenty-five years.

Whereas Saza Chief's family had disappeared without a trace at the time of his death, it was an unprecedented contribution to the study of gorilla behavior that we were able to keep track of young Reuben's mother and the other four females. The five widows wandered about aimlessly like lost souls, in search of a new husband and protector. They seemed to be in a hurry; it was hard to keep up with them. It was ten days before Reuben could report that they had joined a family well known to us, consisting of father, mother and one child. Father, in the prime of life, had become overnight the pasha of a harem of six. For some time, this was a restless troop, always on the move, but eventually they settled down, and a new baby was born.

Only one of the five widows, a youngish one and definitely not little Reuben's mother, gave the impression of not being satisfied with the new arrangement. She became exceedingly aggressive. Gorilla females as a rule are shy and timid and we were puzzled to know what had brought about the sudden change. Did she feel neglected by the pasha?

One day she jumped at Reuben and tried to strangle him. Why? She had met him often and knew he was harmless. He struck at her with his panga. In return she bit him on his hand. He fell on his back and she fell on top of him. He freed himself from her embrace with a well-aimed kick, sending the "nasty bitch," as he called her, back into the bush. The pasha watched the scene without interfering, without any reaction. He certainly was not angry. Was he glad that Reuben had taught her a lesson? She did not learn anything from the kick, however. Two days later she tried to repeat the strangling act. This time Reuben was on his guard and gave her no chance for another embrace.

Reuben delightedly acted out the scene for me. Apparently he had rather enjoyed the experience. I treated his bite with my universal *dawa*—iodine—and, to be on the safe side, had the local dispenser give him a shot of penicillin.

19

The Last of the Kisoro Gorillas

Leopards always have been suspected of being the gorilla's enemy. No reliable observer had proved their guilt, however, and as for me, I believed the theory to be a slander of the big cat. Ironically, it fell to me to be perhaps the first observer to confirm the accusation. I still believe that leopards seldom kill gorillas, but I also know that in our range one particularly huge black leopard did kill many of them.

In 1961 Reuben found two dead gorillas within three days. Both bodies were fresh enough to be given to Makerere Medical School. One of them was kept on ice there for six months and thoroughly studied, and several important papers were written about the findings.

I was deeply shocked, however, when Reuben told me the circumstances of his discovery. At first he had been attracted by the excited behavior of some birds circling over a wooded spot. As he and his trackers neared the place, they heard behind a bush noises such as leopards make, and they had surprised a huge dark cat in the act of killing a red duiker. When the beast noticed them, it disappeared. The little duiker, lying there in its own blood, was still breathing. Near it lay a dead gorilla. A fight obviously had just taken place. Reuben followed a trail of crushed vegetation uphill to a spot where it was evident that a gorilla family had spent the night. The victim, the father, must have been taken unaware while still in bed, and he and his attacker had tussled down the slope. The leopard probably jumped at the unlucky duiker, who happened to pass by and was too easy a prey to let go.

Two days later Reuben came across a second dead gorilla, a young female. He actually saw the killer feeding off the corpse. The leopard made off swiftly, but not before Reuben had had a good view of it. This second body was as fresh as the first one except that the internal parts—heart, stomach, kidneys and liver, mere tidbits to the hunter—had been eaten. The body showed the same gash in the groin as the first one. In both cases the sexual organs had been torn off and an artery severed, which, according to medical opinion, had resulted in instant death.

Thereafter, at intervals of several months, gorilla bodies were found, sometimes partly fresh, sometimes partly decomposed. Pug marks around the vicinity clearly betrayed the identity of the killer. We tried hard to punish him, but hunting leopards in dense undergrowth is tricky. The cats usually sleep by day and hunt by night or in the early morning. Stalking a black leopard in a midnight forest would have been sheer suicide.

Was the killer really a black leopard? The natives called it a "tiger," something between a lion and a leopard. Reuben was quite sure that it was pitch black. There are no tigers in Africa, of course, and hybrids of lion and leopard are also unknown. Was it a melanistic form, a true black panther like those of India, which also surpass the ordinary leopard in size? Or was it only a very dark spotted variation, such as is often found in high African mountains like the Ruwenzori and the Aberdares in Kenya? We were never able to make a clear identification.

Two American zoologists, at different times and independent of one other, confirmed Reuben's statement that our killer was a true black leopard of unusually large size. One, a professor from New

Mexico, watched the leopard stalking a sleeping gorilla during his noonday siesta, a time when these nocturnal hunters are usually in their lairs. The gorilla was lying on his back with his hands crossed under his neck. He seemed to be fast asleep, but he sensed approaching danger and made off leisurely uphill beyond the hunter's reach.

News of our black leopard attracted eager European hunters to our woods. I had the sanction of the Game Warden to let them try to shoot the marauder. An African game warden also stayed at my camp for several weeks at that time and we frequently went out together to stalk the black beast. Every time we were at its heels, it cleverly escaped across the border down into Rwanda, and we could not follow it.

The Rwanda sector of the volcanoes was still a Belgian responsibility at that time. The Belgian concept of conservation does not permit any human interference. "Nature looks after itself," the protectors insisted. "Let the leopard go on destroying gorillas. One day he will meet his match in the shape of a strong male gorilla who will punish him!"

Unfortunately, things did not work out that way. The leopard carried on his bloody business for several years, taking a heavy toll of our gorilla population. I tried a number of times, through influential Belgian friends, to get permission to hunt the killer on Rwanda territory. But the park authorities in Brussels said: *"Non! Non! Non!"* So our gorillas were doomed.

Shortly before the independence of Rwanda, the last of the Belgian game wardens there said to me, "In a few months' time we Belgians will be pulling out of Rwanda. Let's hunt this bloody leopard together on both sides without permission from Brussels." However, before our plans could be carried out, the man was shot dead in his headquarters in Kagera National Park by Watutsi raiders.

Dark or black, this leopard was a ruthless and skillful hunter. His presence in our otherwise peaceful sanctuary made the gorillas shy and wary. We met them less frequently and when we did, the males, instead of staging their mock attack, showed signs of true aggressiveness. Toward the last, we knew of only one group, a family of four with one baby. When that group dwindled to two, I thought: "This is the end of our sanctuary." Naturally I blamed the leopard for our gorillas' annihilation, but I also blamed the Belgian park authorities in Brussels, who had not allowed us to follow the killer into Rwanda.

Reuben still stubbornly patrolled his range, though there were no signs of gorillas anywhere in the area. Then, one day, he reported having seen a group of twelve on the slopes of Mt. Sabinyo. That was good news. I went up myself the next day and saw them. A day later Reuben took a visitor up. The man was intensely interested in seeing the apes, so Reuben let himself be persuaded to follow the spoor down into Rwanda. As he neared his quarry, he heard strange noises, a different sort of barking than that made by gorillas. He hid behind a bush and, thus protected, he watched a party of Batwa hunters, armed with spears, bows, and arrows, and accompanied by their dogs, fall upon those twelve gorillas. As an unarmed intruder on foreign territory, Reuben dared not interfere. He wisely retreated without letting himself be seen. A few days later he found another dead gorilla, a young male with unmistakable spear wounds in his chest.

The Belgians had pulled out by then, and the tribesmen had a free hand in this once ever so strictly protected part of the Albert Park. Why did they hunt apes if they did not eat them? They were poaching all the other game for the pot, and the gorillas would often spoil a day's tracking by coming between the hunters and the hunted. No hunter likes to break off a chase after having followed his quarry, a duiker or bushbuck, for many hours. "Let's get rid of these mischief-makers," they must have thought. That was understandable—but tragic.

The poaching of game and killing of gorillas is still going on in the Virunga forests, in the Rwanda sector as well as the Congo sector.

Recently five gorillas were murdered on the Rwanda side, and others met the same fate in the Congo. The remains showed clearly that the animals had been killed by spear, panga, or arrows. Parts of the bodies were missing. Ears, fingertips, and genitals had been removed, bones crushed and sawed off, indicating that these poor creatures had been killed for juju.

A few African game wardens and rangers have learned to appreciate their wildlife and are doing their best to protect it. I do not want to belittle their efforts. I doubt, however, that the Virunga gorillas will live out the rest of this century.

The leopard eventually disappeared. Perhaps the Belgians were right and he did meet his match in the shape of a strong silverback. But it may also be that the surviving gorillas were too few to justify his efforts and he moved to oew gorilla hunting grounds.

Jill Donisthorpe did not believe me when I wrote her that there were no gorillas left in our mountains. Reuben and I, she suggested, were just being too lazy to leave the beaten track and look for them in less accessible parts. She organized an expedition of ten people and they searched for gorillas for ten days. The party split into three groups. They slept on the tops of the mountains, in the craters, in the forest and swamps. Whenever Jill came down I looked anxiously for a triumphant smile. However, the expedition proved to be a wasted effort. I was depressed to learn that our gorillas were really gone.

During my last years in Kisoro a few gorillas were seen now and then. Our excursions had ceased altogether, since I could no longer afford to pay Reuben and his helpers. Fortunately, my faithful and enthusiastic guide and friend had been hired by the Game Department and was made general game guard of the area.

Whenever gorillas were seen in our sanctuary after that, the rumor would spread and guests would come to me and say, "We have heard your gorillas are back."

"Back? From where?" I would ask. "Back from 'the undiscover'd country from whose bourn no traveller returns'?"

20

Dian Fossey: Friend of the Apes

No one knows gorillas as well as Dian Fossey. The first time I met her she was visiting Kisoro as an American tourist. She had seen some of our gorillas and had met some in the Congo also. She had always wanted to return for further study in order to write a children's book about gorillas.

In 1967, several years after the disappearance of the gorillas from our sanctuary and after Independence came to Uganda, Rwanda, and the Congo, I was surprised and delighted to welcome Dian Fossey once again to Travellers Rest. This time she was accompanied by Alan Root, the naturalist and wildlife photographer from Nairobi. When they arrived, life in free Rwanda and Uganda was safe and peaceful, but the Congo, in the seventh year of its independence, was still pandemonium. I was shocked and dismayed when the two told me that they were on their way to that country, where Dian, sponsored by Dr. Leakey and financed by the National Geographic Society, was to observe gorillas. Alan was taking her as far as Kabara, to Schaller's old camping site, to help her get settled. She was then to stay up there for two years with only two or three native helpers.

"That is madness!" I said. "I cannot understand Dr. Leakey sending this woman to the Congo at a time when most of the whites have fled. Are you really so ignorant in Nairobi about what is going on in the Congo?" I tried hard to dissuade Dian and Alan, but nothing would deter them. On his return a week later, Alan reported that all had gone well and things had seemed quiet. They had had no difficulties at the border and Dian had settled down happily.

The thought of Dian being alone in the Congo continued to worry me. What could Dr. Leakey have been thinking of? How had he persuaded Dian to expose herself to such danger?

Later, I learned that Dian had attended one of Dr. Leakey's lectures back in America. He had spoken of the importance of studying gorilla behavior in its natural setting. Referring to Schaller's comprehensive monograph, he had pointed out that there were still many gaps to be filled. He was looking, he said, for a suitable person to take on that job. After the lecture, Dian had gone to him and said, "Dr. Leakey, I am the person!"

"Which person?" he asked.

"The one who is going to the Congo to study gorillas."

Dian, an occupational therapist by profession, had no training for the work that Dr. Leakey had in mind. It was by a stroke of genius that he chose her in preference to many well-trained students who also had applied. He realized that formal academic training was less important than the qualities of character he saw in Dian. He could not have chosen more wisely.

Dian was to come once a month to Kisoro to collect her mail, replenish her larder, and to have a good talk and a hot bath at Travellers Rest. I awaited her first visit with some anxiety, but I need not have worried. She was fine, she declared. She had been well treated at the border. She had already made contact with some groups of gorillas and was perfectly happy.

She seemed a little less enthusiastic the second time she came to the inn. She had been stopped on the road by a soldier who had demanded to see *le reçu*—the receipt—for some goods she had bought in Goma. Dian could not understand what he wanted and he enjoyed her embarrassment. He had repeated his demand again and again and had driven her almost to desperation before she eventually understood. It was a minor incident but it illustrates the early difficulties Dian encountered because she did not know any of the languages spoken in the area.

Shortly before she was about to start back the next morning she doubled over with a painful attack of stomach cramps.

"Aha!" I thought. "This is a psychological case. Dian is fleeing into illness so that she will not have to return to the Congo!"

I was mistaken. Dian was not faking illness. She stayed one more day and left the following morning in the best of spirits. I stopped worrying.

Dian was very successful with the apes. They began to accept her as a part of their surroundings and she was able to add many new facets of their life and behavior to Schaller's observations. I had to admit that I had been wrong and Dr. Leakey right: "It pays to take risks in the interest of science!"

Dian is not an expert at climbing trees. One day a group of gorillas watched her try to get hold of a low branch to pull herself up to a vantage point.

As she clung to the branch, intermittently dangling and falling back to the ground, the gorillas seemed to be enjoying her performance. For the first time, instead of fleeing, they gave way to their curiosity and approached within twenty feet to sit and stare at her clumsiness. To humor them, Dian prolonged her struggle and let herself fall several times. No wonder the gorillas began to like her and lose their fear of the strange, awkward creature.

But life was not all fun up in Dian's camp. It was hard to climb up and down steep slopes in every kind of wind and weather, to scramble through wet ground vegetation, to be stung by nettles and scratched by thorns and thistles. Sometimes Dian had to drive off native poachers, and Watutsi herdsmen with their cattle, who had no right to be there. On one occasion poachers illegally killed a buffalo and calf. The park guards were able to confiscate the meat and invited their friends from lower down to join the feast. The eating, drinking, and gaiety went on around her tent for some days and that was a trying experience, but the natives respected her and made no attempt to touch her.

The peaceful interlude did not last long. The Congo burst out again, and the ensuing events justified my forebodings. I began to worry. Was she safe there in the mountains? Would the warring factions leave her undisturbed, allowing her to continue her work?

Soon a gang of porters arrived, took down her tent, and brought her and her equipment back to her base in Rumangabo, headquarters of the Albert National Park. To all intents and purposes she was put under house arrest. And she had no idea why.

On returning from the mountains, Dian had received a letter from Customs informing her that the permit on her Landrover had expired and that she must come to the border to renew it and pay duty. When she reached the border she found the entire customs staff very drunk indeed. When they discovered that she did not have the cash they demanded, they threatened to beat her up and confiscate the Landrover. Dian, refusing to be intimidated, put the engine key in her breast pocket, crossed her arms defiantly and said: "Come and get it, if you dare!" The act had a sobering effect on the men and they let her go on to Kisoro to collect the money they demanded.

Dian did not arrive in good spirits on that occasion. She was disgusted and fed up with the whole Congo. She telegraphed Dr. Leakey for instructions, mentioning the exorbitant amount that the Congolese Customs wanted for the permit.

The next day the men at the border, having sobered up, sent an apology and said that the American lady could come back. No harm would be done to her if she paid the required duty.

Dian could not resist the call of her apes and announced that she was going back. I tried to dissuade her, but when I failed to do so, I borrowed the money she needed from my Indian friend, Mr. Lalji, and I went with her to the barrier, where all went smoothly.

Shortly after Dian's departure, an airplane circled over Kisoro. Dr. Leakey, responding to Dian's telegram, had sent the pilot to rescue her. The plane had to return without Dian.

The next time Dian came to Travellers Rest she was accompanied by a young, unpleasant-looking Congolese man. Her excitement was at fever pitch, her throat choked with anger and indignation. It was some time before she could speak coherently.

On her return to Rumangabo everything had seemed all right at first and she had made arrangements for the safari back to Kabara. Just before leaving, however, she was suddenly informed that the Army would not let her to go; she would have to remain, until further notice, at her base in Rumangabo. Incensed, Dian went to the nearby garrison of the Force Publique, in the hope of persuading the C.O. to let her return to Kabara. Dian's reception in this lion's den of drunken soldiers was by no means friendly and the permission she sought was refused. After a few more days under house arrest, she received a notice asking her to pay an additional amount for the Landrover. This again gave her an excuse to ask permission to come to Kisoro to raise the money. That she succeeded in keeping the drunken rabble at the border at arm's length and withstood their threats of violence, talking them into letting her go, was a victory of the spirit over brute force.

The sordid individual with Dian was responsible for bringing her back. "Miss Fossey is not your prisoner," I said. "She is going to stay here. If she wants to return, I shall tie her to that tree out there!"

"But I have guaranteed to bring her back," her guard protested. "They will shoot me if I don't."

"Better to shoot you than her," I said. Eventually he left.

Dian was miserable at the thought of giving up the work which meant so much to her and which already had shown promising results. What should she do? Wait for "good weather" in the Congo? Return to America? I suggested that she make a survey of the gorilla situation in the Rwanda sector. I pointed out that little was known about that area; it had been neglected even by the Belgians. We knew that much poaching went on over there, and a number of gorillas had been killed since Independence. Dian's presence, I suggested, might serve to intimidate the hunters and thus help protect the animals.

I told Dian about Madame de M., a Belgian, who had a pyrethrum plantation on the slopes of Mt. Karisimbi. That might be a good starting point for such investigations, I said. Dian and Madame de M. would make a good team. Like Dian, Madame de M. was an adventurous soul. An expert on our two active volcanoes, she had once owned a snake farm, also had once walked for many weeks through the Ituri Forest to meet the unspoiled Pygmies. She had recently lost her husband and needed company. When I approached her with my suggestion, she readily agreed to let Dian use the plantation as her base camp.

Dian flew to Nairobi, discussed the new project with Dr. Leakey, and soon returned to Rwanda. She arrived at Madame de M.'s plantation soon after her hostess had suffered a second personal tragedy. To divert Madame's mind, Dian interested her in the gorilla investigation. The two went out together, camping in the forest and searching the slopes of Karisimbi for the apes.

The investigation yielded poor results. However, when Dian moved her camp to the south side of Mt. Visoke, she discovered a veritable gorilla paradise. There, the dream of her life came true. Her efforts to make friends with gorillas were blessed beyond her wildest expectations.

The gorillas had been driven into this region by the poachers and herdsmen, who had taken possession of the once protected area after Independence in Rwanda. Only a resourceful woman like Dian would have thought of using Halloween masks to frighten off these human intruders. She found a particularly gruesome contraption, designed to be worn in the mouth, with long, frightening canines protruding. To test its effect on Africans, she held a dress rehearsal at a tea party. The native servant, who a few moments earlier had witnessed Dian as an attractive woman, returned with the tea and saw a monstrous creature grinning at him. He dropped the tray and ran away screaming. The masks proved just as effective with the poachers and herdsmen. For some time at least they believed she was a white witch.

Tough and tenacious, Dian visited her gorilla groups almost daily, even in the worst weather. Working with infinite patience, she succeeded in gaining their confidence. Some of them eventually accepted her almost as a member of the family. The younger apes, in particular, soon overcame all fear and shyness. They would come very close, examine her camera, and fiddle with her bootlaces. One young male was so fascinated by her gloves that he inspected them finger by finger and sniffed them. When his curiosity was fully satisfied, he dropped then nonchalantly to the ground. Dian had friends also among the elders of the groups. Her favorite was a hefty silverback, a colossus weighing about 400 pounds. She called him Rafiki, which in Swahili means friend. Rafiki was the leader of a group of five males, a kind of bachelors' club. A group with no female members is very unusual in the social organization of these apes. Rafiki's five bachelors were childless, but when Dian first met them their lives centered around a senile dowager. The affection the male apes showed for the old lady was touching. Dian used to watch her and Rafiki embracing each other. They often shared the same bed at night. One day the two disappeared, and a couple of days later Rafiki returned alone. Had the old one felt that her end was near? Had she wished to die away from the others, with only Rafiki beside her?

"How old do you think this gorilla matron was?" I asked Dian, when I heard the story. Dian replied that she had come to the conclusion that Koko defied all accepted estimates of her age. To judge from her apparent senility, her baldness, and her wrinkles, Dian guessed that she might have been around fifty years old. No one has yet observed the natural life of a gorilla from birth to death. It is thought that twenty, perhaps thirty, years is the span of a gorilla's life in the wild. Do they live to be older in captivity? It would seem so. Bamboo in the Philadelphia Zoo was thirty-five when he died; his successor Massa—now the doyen of all zoo gorillas—is approaching forty and London's Guy is well over twenty-five.

I once saw an amazing photograph of Dian and Rafiki. He, with crossed arms, leans against a fallen tree; she stands with her arms folded only a few yards away. Among gorillas, folding the arms is a sign of submission. The picture is deeply moving: the two, the human and the ape, seem to be engaged in a profound conversation.

Dian was not content to watch gorillas unobtrusively, to be merely tolerated, to wait for them to make the first step toward a cordial relationship. "In Rome one should do as the Romans do," she thought. "Why not do in gorillaland as the gorillas do?"

Dian practiced gorilla behavior and soon became very adept. She learned how to bark, grunt, and roar, how to beat her chest, and how to eat their food. Usually it was only necessary for her to pretend to eat gorilla fare but there were occasions when the suspicious apes watched her so closely and for so long that she had to swallow the stuff, leaves, stalks, and all!

When the animals were in a nervous and aggressive mood, Dian would sit down, pretending to be completely unconcerned, interested only in munching the stalks of celery in her hand. Once, before she and Rafiki knew each other very well, he and his companions charged her. Rafiki stopped about three feet away. Dian realized she couldn't rely on the feeding stunt. She had to prove that she could demonstrate just as well as they. She got up, spread her arms, and roared! Rafiki halted and retreated.

Later, she learned that the cause of their tension that day was the presence of a second group of gorillas above them and a lone silverback below them. The noise of Dian's approach through the thick foliage gave Rafiki and his companions an excuse to charge at this unknown third distraction and relieve their nervousness.

144

Once a gorilla youngster, practicing tree-climbing near where Dian sat, managed to pull the tree down. The noise alarmed his parents and brought them rushing to the spot. They seemed to suspect that Dian had caused the accident. Fortunately, their tension was diverted when an even smaller child climbed into the fallen tree and was rescued from a perilous position by the first acrobat.

Dian's field work was interrupted when she became foster mother to two gorilla infants. Coco, a male about sixteen months old, and Pucker Puss, a female about two years old, had been captured by two inexperienced natives, by order of the Rwanda Government. The Minister of Information had promised them to his friend, the Mayor of Cologne, for the latter's zoo.

Dian sacrificed more than two months of her own work to nurse the frightened, half-starved infants back to health. They would have died without her attention. In caring for them, she learned much about gorilla behavior and feeding habits. She used to take her youngsters into the forest and let them collect their own food. She enjoyed their company and was sad when they left her for the long journey to their new home in West Germany.

The two young apes are now well looked after in the Cologne Zoo, fortunately, although this does not excuse the fact that their two family groups had to be slaughtered in order to obtain them for public exhibit. The zoo experts, however, were disappointed when the Olympic Test (microscopic examination of hair, the chromosomes of which reveal sex) proved that both babies were female. International wildlife conservationists were indignant when they heard of the capture of the two small apes. Professor Grzimek, the protector of Africa's wildlife, demanded that they be returned to their native mountains, which, of course, was never done.

Yet, sadly, it is possible that today the chance for survival of the species is better in captivity than in the wild. There are now over 300 gorillas living in zoos and research centers. About 30 of them have given birth to healthy offspring, and in some places a third generation is growing up.

The recent census that Dian undertook with the help of some

Cambridge University students has established that the gorilla population of the entire Virunga range has dwindled to roughly 300 from the 800 estimated in 1950. More critical for the survival of the species is the inexplicable fact that males now outnumber females two to one, whereas previous studies demonstrated the opposite. As females are more important to the survival of a species, the mountain gorilla seems doomed to extinction.

Dian's findings have upset some previously held ideas about gorilla behavior. I had believed gorillas did not strive for supremacy within their group. But Uncle Bert, a silverback that Dian observed, did not hesitate to eject a second male and make him a solitary errant when taking power from a deceased leader. Apparently he wanted to rid himself once and for all of any rivalry. Furthermore, I had believed that gorillas do not covet their neighbors' wives. But Dian told me that Rafiki and his club of bachelors, evidently tired of living without females, deliberately followed Uncle Bert's group for weeks. Eventually, Rafiki abducted one of Uncle Bert's harem. Uncle Bert did not accept it with good grace and gave chase. They came to blows, resulting in rather severe injuries on both sides, but Rafiki not only kept the lady he had abducted, he even kidnapped a few more young females. Soon the ex-bachelors' club was no longer childless.

In January 1970 Dian interrupted her observations to journey to Cambridge University to analyze her field notes and prepare her doctor's thesis based on her work. She wrote me from England:

> During the last week before leaving I had some of my best contacts with Group Eight. On the very last day, Peanuts, the young blackback, came up to sit by my side for some twenty minutes and rapped my hand twice before leaving me! You can well imagine what this meant to me after all this time! It was the best going-away present I could have had, but naturally it made it all the harder to leave.

In a recent letter Dian reports her amazing progress:

The majority of the study animals in the four different groups have learned to accept me and my student observers in a totally trusting manner. Without any "threat and reward" kind of stimulus, various gorilla individuals, especially the young, now seek me out just to investigate my clothing, the contents of my knapsack and to play with my extremities or pull my hair. They seem fascinated with the proximity of a human they can trust, and often build their day nests adjoining the spot where I have settled down to make contact with the group.

Once Dian shared the same tree with a hefty young gorilla male. On climbing down he put his hands gently on her shoulder and pushed her—not quite so gently—out of his way. No harm was meant. He merely thought she was too slow in getting down.

On another occasion, Dian was sitting on a fallen tree trunk when Peanuts, her best friend, came and sat down beside her. She happened to have with her a little pygeum, a fruit of which gorillas are very fond. She put three berries on her palm and held it out to Peanuts, who gratefully picked them up one by one, clearly disappointed that there weren't more of them.

Dian was the first person to learn that gorillas are not only willing to shake your hand, but to eat out of it!

21

Farewell to Africa

It was reluctantly, and with a heavy heart, that I came to the conclusion in 1969 that the white man's days in black Africa were numbered and that I had better pull out while the going was still reasonably good. Independence made me a foreigner in Uganda, merely tolerated, liable to be expelled at any time it suited the authorities. I no longer felt at ease. All my life I have been on the side of the underdog—and in Africa on the side of the Africans. Now the winds of change blowing over Africa had made me the underdog and my life uncomfortable.

In retrospect, I realize that it was the disappearance of the apes from the mountains that first took the glamor out of my existence. The gorillas, and the visitors who came to see them, had daily given new stimulus to my life. Without them, life had become "stale, flat and unprofitable."

My democratic disposition did not approve of President Obote's one-party system; it had the scent of fascism, to which I was allergic. I could see it turning into a military dictatorship and I had had enough of soldiers and policemen ever since, during the Congo turmoil, Kisoro had been occupied by the Uganda army. The wild, unruly soldiery had been a source of constant irritation and had made itself hated and despised by the local population. In order to protect us against a (most unlikely) invasion from the rebellious Congo, our protectors had closed the Congo and Rwanda borders, virtually cutting off Travellers Rest.

Visitors could no longer come over to the inn. Many were molested, even arrested as spies and put into prison. My business came to a complete standstill and I was driven to the verge of bankruptcy. But my morale, I found, had suffered even more than my bank account, and this was a loss that no spell of good business could repair, not even that of the Indian traders, who now had turned Travellers Rest into a smugglers' den from which they conducted their shady business with the lawless Congo. Like an American gangster film, hostile racketeers sat in corners, hats shoved back, wrapped in clouds of smoke, drinking gin and playing poker, waiting for cars coming from the Congo. According to the natives, there was much swapping of tires in which contraband gold and diamonds were hidden. The Congo cars returned and the Uganda ones—with alternate tires—went unchecked through the Customs Post on to Kampala. At times long convoys of lorries loaded with Kivu coffee crossed our border with forged documents in transit to Mombasa and on to the European markets. It was a complicated business, but fortunes were made in no time. Everybody thought that I had my fingers in the gold, diamonds, and coffee that changed hands at Travellers Rest. I lacked the courage for that line of business, even if I had wanted to engage in it.

The senseless murder of Madame de M.'s son and his friends in the Congo confirmed my resolution to leave Africa. Madame had told me that the three students from Louvain University were on the way to her plantation and that they would spend the last night of their journey at my inn. In due course the three Belgian students arrived and we had a happy party with some congenial British students who were lodging there.

The next morning the three Belgians shook hands and headed for Rwanda (so I thought). Some days later, Madame arrived at the inn—alone. An indescribable foreboding of disaster seized me and I could hardly get out the words, "What brings you here, Madame?"

"I am looking for my boys," she answered. "They should have arrived days ago. Maybe they have had some trouble with their old Landrover and I can help them."

It was one of the most awful moments of my life. I had to tell her they had been at my inn three nights before. We both knew they must not have reached Rwanda; one heard of anything that happened to a white person there. The boys must have taken the wrong road and driven to the Congo. At the Congo border, people confirmed that three young men had crossed over a few days before and been arrested by the soldiers as Belgian spies and mercenaries. Later we heard they had been tortured and cruelly killed.

We all hear daily of greater tragedies, but the deaths of these boys concerned me personally. They had spent the last night of their lives under my roof. I had been the last person to shake hands with them. Suddenly, I felt that I had had my fill of Africa.

In the end, however, it was a trivial incident that made up my mind. I was forced to hang the portrait of President Milton Apollo Obote, whom I so thoroughly disliked, upon the wall of Travellers Rest.

Perhaps my response was unwise, but I had never lived in a country where the showing of the Führer's picture was required. The British had never interfered with my interior decor and during the first six years of Independence I had never hung Obote's portrait. When reproached, I would explain that such portraits were not my style; my two large pictures of gorillas were more in keeping with the atmosphere of the inn.

At last a drunken, corrupt assistant district commissioner commanded me to hang the hated portrait in my lounge. I was outraged that I was no longer master of my own house. I was tired of having to cater to, must less to fear anyone. I wanted to be free again.

At that point I did not hesitate to sell Travellers Rest to a Swiss, who made what seemed a fair offer: one-third cash, the rest in shares in a big hotel he was building in Kigali, capital of Rwanda. Anyway, I had no choice; there was no other buyer.

The Swiss had grand ideas—a chain of tourist lodges on the Uganda side of the border. Travellers Rest was to be the starting point for safaris to the country's game parks. He added four rooms to the inn and the first parties arrived while I was still in Kisoro.

Alas, the capital that had been promised him failed to materialize. Shortly after I left Uganda the Swiss had to liquidate his entire African interests and my shares were a total loss. Disastrous as it was, it turned out to be a blessing in disguise. For had I stayed on, I would have been expelled by Amin's government and left the country almost penniless, which was what happened to the Indian to whom the Swiss sold Travellers Rest. He was in the process of adding a new wing to the inn a few weeks before he was ordered to hand the place over to the government without compensation and to leave the country within three weeks, with only a pittance in his pocket. Travellers Rest is now run as a part of Uganda Hotels, the concern that owns all hotels in the country.

It was painful to leave Kisoro. I had wanted to spend the last years of my life there, in a hut—round or square—with gentle Cosmas, my waiter and friend, looking after me. And now, in my late sixties, I had to leave my sanctuary and look for another niche.

I left Kisoro at four one morning. A friend and Peter, a young African who had adopted me as a father and become my right-hand man, took me to Kabale, where I was to catch the bus to Fort Portal. All the men were up to say good-bye, even those from Nyarusiza. Rex and Sombra, the dogs, were roaming the garden; it was easier to leave without seeing them. It was still dark when we reached the Kanaba Gap from whence I had once looked down on the Kisoro plain in anticipation. Now I did not look back and could not have, even if the sun had been up.

I retraced my route of fifteen years before, traveling via Kampala and Nairobi to Mombasa. My ship sailed at midnight on February

28, 1969, and my papers did not allow me to stay one minute longer. I left Kilindini Harbor with a sigh of relief. I did not look back to the vanishing lights of Mombasa, not back to the coast of East Africa. Kisoro—Uganda—East Africa: There had been my life. Now it was all over; there was no room for me in the new Africa that was emerging.

I am hopelessly old-fashioned, a disillusioned romantic, as someone once called me. I believe in the patriarchal way of life so scorned by all fanatical anticolonialists. I think colonial rule had its good points and probably suited the Ugandans much better than the present dictatorship, where they live under the constant fear of being robbed, murdered, or imprisoned. Certainly life was safer there in the past.

I loved being a father to my people, enjoyed the confidence they placed in me. With pride I think of some of my protégés who are doing well today. Charles, who was so eager to learn, became a high school teacher; Peter is now a manager of a hotel in the Ugandan chain; and David, an able agriculturalist, developed a fine farm of his own under an irrigation scheme. With warm sympathy and gratitude do I think of all my helpers: of Saga, the jolly cook; of gentle Cosmas, the headwaiter; of Elias, the clever carpenter; and, above all, of Reuben, my guide, and his splendid men. I think too of my willing and hard-working gardeners and of Sawa Sawa, the beggar clown, whose portrait looks down on me as I write. I used to receive many letters from African friends, who now, for fear of the censor, do not dare to write. All started, "Dear beloved father," and ended, "Your loving son." Are not such close personal relationships between human beings worth infinitely more than a politician's portrait hanging on the wall?

I think, too, of my guests, both great and small, many of whom were pioneering in Africa, and some of whom became my friends. I think of the humble beginnings of our gorilla searches, how they stimulated scientific investigation and led to the ongoing work of Dian Fossey under the National Geographic Society. And when I think of that magnificent—unforgettable—landscape I feel home-sick for Uganda!

Where else on earth could I have felt that I was lord of three mighty volcanoes, of forests and ravines where, side by side, lived the legendary gorilla, with elephant, buffalo, leopard, and all kinds of other wonderful creatures? Sitting up there in my camp at 10,000 feet, looking down over the wide plain, over hills, valleys and lakes, I was always strangely elated. I often felt—yes, really—as if I were king of the fabled land of the gorillas.

Epilogue

Winds of Change
Sweep Travellers Rest

The Swiss who took over Travellers Rest when I left was ambitious. He envisioned a great Safari Lodge, attracting a growing tourist trade, and he was confident that he had the secure financial backing to make his dream come true. He set about modernizing and expanding the inn. He added a stable of saddle horses. And, because he had the foresight to realize that in the new Uganda the visible management of his lodge must be native, he arranged for two promising young men to go to Switzerland for thorough training in the profession of hotel management.

I was kept in touch with his progress through letters from Dian Fossey. She continued to visit Travellers Rest whenever she came to Kisoro for supplies, and the letters in which she reported on these visits were touched with nostalgia. Soon after I left, she wrote:

> I don't know how to tell you how much I miss my second home in Kisoro. . . . You, the silly boys, my Rex, my room are part of the reason that I insist upon continuing my work in this area.

Her special feeling for the place made her understand that my yearning for news of Travellers Rest was not limited to curiosity about physical changes and management innovations. She wrote of these, but she also wrote of the dogs, the gardens, the men who had been my helpers, and my friends.

Excerpts from her letters between 1969 and 1971 show not only how completely she felt at home at Travellers Rest but also how reluctant she was to accept the changes taking place:

March 17, 1969

I have only been to Kisoro once—on the 24th of February— but there were no guests that night except for Africans. . . . Your men told me they had been very busy. I must say Peter certainly does an efficient job. . . . The others all insist that you will return but Sawa Sawa knows better. That old man really does love you, Walter. The dogs seemed fine and naturally both slept in my room, but it wasn't the same. . . .

April 18, 1969

I spent last night in Kisoro and showed your letter to Sawa Sawa just to point out his name and your message of love to him, and, Walter, he really did understand. . . . Peter is still trying to do well. . . . The place was overbooked, with tents on the lawn, and I was moved to your little room since people were still pouring in. They were mainly the "peace corps" type. . . .

Last night the cook and I and Cosmos did dishes until 11:30, and it was fun because they were good company and we all talked about the "old days.". . . The garden looks fine as do the dogs and cats. As you know, they are building an extension of four rooms on the lawn area that faces the road toward

Bunagana. Thus far only the lawn is dug up and blocked out, but all the building blocks . . . are piled up on the rest of the lawn.

By autumn of that year, Dian's uneasiness had increased. She wrote that the inn seemed "dismal," even though she did enjoy "saddling up one of the nags that Roland calls horses and plodding around on the back roads." The same letter indicated the obvious reason for the gloomy atmosphere:

September 10, 1969

This particular visit seemed rather pathetic as most of the men have been fired, including Peter, and the two that remain—Cosmos and little Pascal—plus one other whose name I don't know—weren't very happy. The newly fired ones came to my room with long, sad tales and all wanted to write to you but only Pascal and Cosmos brought me letters to send. . . . Both of the dogs are very fit and one of the cats is pregnant again. . . . I am afraid the food is really *bad*—the only thing the new cook can't ruin are the pieces of fish. . . .

At that time Dian estimated that the new building units were almost completed and that they would be ready for occupancy "at the end of two more months." Before those two months had passed, however, disaster had overtaken the Swiss entrepreneur. His backers in Switzerland had soured on the prospects for hotel development in East Africa and no more money was available. He was forced to liquidate his interests. Travellers Rest changed hands again, and in the process of the transfer I lost the money I had so hopefully invested in the project.

The new owner was Indian, and with his coming a new day seemed to dawn. Visiting Travellers Rest in December, Dian was jubilant about the change in atmosphere:

Roland has gone and Mr. Jamboni is holding down the fort.
. . . The men like him and the whole feeling of the place is
happy again. . . .

The food is much better now and everything is clean including the garden. . . . Still no electricity so I made myself
pretty unpopular by taking a light from the bar, but I'll be
damned if I'll pay 66 shillings (without a drink) for dinner and
breakfast and not have a light in my room.

The horses are gone—some Indian who works in the
forestry department bought them for almost nothing. I wish I
could have had first choice!

The same optimism was apparent in Dian's letter in January
1970. The boys, she said, all liked the new owner and even the dog
Rex, who had been "just a rib cage" when the Swiss owner left, was
in much better shape.

Six months later a long report detailed the improvements—
including hot water and electricity—made by Mr. Jamboni. Dian
was enthusiastic about the future of Travellers Rest:

June 30, 1970

You know I love Travellers Rest because of what you have
made it mean to me. . . . It will never, of course, be the same
without you, but . . . I am truly grateful to see at last someone
who will take care of it properly. The man's motive may only
stem from financial gain, but still he is giving it his total attention and I do wish him well. . . . Also, the other men seemed
content and happy to be really working again. [They] seemed
proud again for the first time since you left.

Mr. Jamboni undoubtedly was proud, too, as he saw his hard
work bearing fruit. Travellers Rest was certain to become a hostelry

drawing tourists from all over the world. Then, suddenly, the whole enterprise came crashing down. Idi Amin the Great decreed a purge of Uganda's Indians. Mr. Jamboni was given thirty days to leave the country. He was permitted to take fifty dollars with him.

Little news of Travellers Rest has come to me since that time. Dian Fossey no longer makes trips to Kisoro. Her last visit at the inn was on July 12, 1972, just before the ouster of Mr. Jamboni. She still writes to me but only about her work. Once in awhile a letter comes from one of my former employees, but they cannot write in English and thus must dictate their communications—very guardedly—to a friend. I do not dare keep up my correspondence with my intellectual African friends for fear the contact with me might embarrass them.

But I have reason to believe that Travellers Rest still exists, and even that some of the same men are working there. As late as last December, a letter from David Rwiseburi reported that Sawa Sawa is still alive, and that Saga, Cosmos, Elias the cook, and the Shamba boys are still working at the inn.

Today, Travellers Rest is a government-owned enterprise, part of the Uganda Hotel chain. It still draws some tourists. Now and then a few gorillas are seen in the vicinity. But I am certain that the air of freedom and camaraderie that marked the old days is gone, perhaps forever.

Index